WITHDRAWN
FROM
UNIVERSITY OF PLYMOUTH
LIBRARY SE…

16.

Scientific, Environmental, and Political Issues in the Circum-Caspian Region

NATO ASI Series

Advanced Science Institutes Series

A Series presenting the results of activities sponsored by the NATO Science Committee, which aims at the dissemination of advanced scientific and technological knowledge, with a view to strengthening links between scientific communities.

The Series is published by an international board of publishers in conjunction with the NATO Scientific Affairs Division

A Life Sciences **B Physics**	Plenum Publishing Corporation London and New York
C Mathematical and Physical Sciences **D Behavioural and Social Sciences** **E Applied Sciences**	Kluwer Academic Publishers Dordrecht, Boston and London
F Computer and Systems Sciences **G Ecological Sciences** **H Cell Biology** **I Global Environmental Change**	Springer-Verlag Berlin, Heidelberg, New York, London, Paris and Tokyo

PARTNERSHIP SUB-SERIES

1. Disarmament Technologies	Kluwer Academic Publishers
2. Environment	Springer-Verlag / Kluwer Academic Publishers
3. High Technology	Kluwer Academic Publishers
4. Science and Technology Policy	Kluwer Academic Publishers
5. Computer Networking	Kluwer Academic Publishers

The Partnership Sub-Series incorporates activities undertaken in collaboration with NATO's Cooperation Partners, the countries of the CIS and Central and Eastern Europe, in Priority Areas of concern to those countries.

NATO-PCO-DATA BASE

The electronic index to the NATO ASI Series provides full bibliographical references (with keywords and/or abstracts) to more than 50000 contributions from international scientists published in all sections of the NATO ASI Series.
Access to the NATO-PCO-DATA BASE is possible in two ways:

– via online FILE 128 (NATO-PCO-DATA BASE) hosted by ESRIN, Via Galileo Galilei, I-00044 Frascati, Italy.

– via CD-ROM "NATO-PCO-DATA BASE" with user-friendly retrieval software in English, French and German (© WTV GmbH and DATAWARE Technologies Inc. 1989).

The CD-ROM can be ordered through any member of the Board of Publishers or through NATO-PCO, Overijse, Belgium.

Series 2: Environment – Vol. 29

Scientific, Environmental, and Political Issues in the Circum-Caspian Region

edited by

Michael H. Glantz

Environmental and Societal Impacts Group,
National Center for Atmospheric Research,*
Boulder, Colorado, U.S.A.

and

Igor S. Zonn

Russian National Committee
for UNEP (UN Environment Programme),
Moscow, Russia

* The National Center for Atmospheric Research is sponsored by the National Science Foundation.

Kluwer Academic Publishers

Dordrecht / Boston / London

Published in cooperation with NATO Scientific Affairs Division

Proceedings of the NATO Advanced Research Workshop on
The Scientific, Environmental, and Political Issues of the Circum-Caspian Region
Moscow, Russia
13–16 May 1996

A C.I.P. Catalogue record for this book is available from the Library of Congress

ISBN 0-7923-4626-2

Published by Kluwer Academic Publishers,
P.O. Box 17, 3300 AA Dordrecht, The Netherlands.

Sold and distributed in the U.S.A. and Canada
by Kluwer Academic Publishers,
101 Philip Drive, Norwell, MA 02061, U.S.A.

In all other countries, sold and distributed
by Kluwer Academic Publishers Group,
P.O. Box 322, 3300 AH Dordrecht, The Netherlands.

Printed on acid-free paper

All Rights Reserved
© 1997 Kluwer Academic Publishers
No part of the material protected by this copyright notice may be reproduced or utilized in any form or by any means, electronic or mechanical, including photocopying, recording or by any information storage and retrieval system, without written permission from the copyright owner.

Printed in the Netherlands

Dedication

This book is dedicated to *Dr. Sergei Zonn*, who has studied desertification in the Black Lands of Kalmykia for most of his professional life. Still active in his ninetieth year, he produced a background paper on Kalmykia for workshop participants.

Table of Contents

FOREWORD

On behalf of the Russian Federation Committee on Water Economy, I would like to welcome the participants of the NATO-sponsored Workshop on the Problems of the Caspian Sea and the Circum-Caspian States.

The world's largest intercontinental sea/lake is well known for its wealth mineral and fuel resources and sturgeon stocks' the products of which (oil and caviar) are in constant demand on the world market.

During the last half-century the Caspian Sea has been the focus of the scientific community concerned with its level fluctuations. We were, and still are, solving a two-faceted issue: to rescue the Caspian Sea and to rescue the population from the Caspian Sea.

To rescue the Caspian Sea is to address a broad spectrum of environmental issues related primarily to water pollution by waste water and petroleum products.

To rescue the population from the Caspian Sea means that an almost 2.5 m sea level rise in the last two decades has resulted in flooding of vast coastal areas deteriorating economic and social spheres of activity.

Solutions to these issues are linked to the numerous mysterious aspects of recent Caspian Sea behavior.

Regrettably, the collapse of the USSR has led to a decline of marine observations and control over the use of marine resources in the region. Coordinated international action on the protection of living marine resources have terminated, generating disastrous consequences.

Today, the Caspian Sea is an international water body with a new balance of political forces following the creation of the newly independent states -- the former Soviet republics. The only way towards sustainable development of the region is for increased integration of activities related to the sea in the Circum-Caspian states.

NICOLAI N. MIKHEYEV
First Deputy Minister
National Resources of the Russian Federation
Moscow, Russia

PREFACE

MICHAEL H. GLANTZ
Environmental and Societal Impacts Group
National Center for Atmospheric Research
Boulder, Colorado USA

IGOR S. ZONN
Russian National Committee
for UNEP (UNEPCOM)
Moscow, Russia

The chapters that follow address a variety of climate- and resource-related environmental issues in the circum-Caspian region. Each contribution was selected to fulfill a specific objective. Contributors were drawn from various academic disciplines and from different countries and international organizations concerned directly with Caspian Sea issues and from those with similar environmental problems.

Nicolai Mikheyev, Deputy Minister for Water Resources, opened the workshop by drawing attention to the importance of improving our understanding of the fluctuating and changing Caspian environment and the sustainable use of its natural resources.

In Part I, *An Introduction to the Caspian Sea Environment*, **Michael Glantz** raises the possibility of fostering regional cooperation based on shared environmental problems, such as those in the Caspian region. **Igor Zonn** presents an overview of the environmental situation in the Caspian Sea basin, as well as the socioeconomic consequences of Caspian Sea level rise. **Genady Golubev** raises the issue of the need for research on the Caspian and other closed (terminal) lakes in the world, while noting that the Caspian is a somewhat unique inland sea with characteristics of both a lake and a sea. Such systems, he notes, are extremely vulnerable to the adverse impacts of human activities.

George Golitsyn, Director of the Institute for Atmospheric Physics, provides a history of Caspian Sea level changes over the past century and a half, addressing some of the hypotheses about the cause of level fluctuations. He also addressed the problems and prospects of forecasting Caspian Sea level changes. **Tatyana Saiko** addresses problems in the Caspian region generated by conflicting national priorities for both oil and the highly valued, caviar-producing sturgeon fishery. **Sergei Vinogradov** presents legal discussions relating to the status in the eyes of international jurists of the Caspian region. Is it a sea or is it a lake? Depending on the resolution of this legal issue, the resources of the Caspian will be decided accordingly. The implications of this distinction are major with regard to control of pollution in the Caspian and with regard to the implications for resources exploitation in and around the sea.

Part II focuses on *The Impacts and Implications of Caspian Sea Level Rise*. In this section several national and sub-national perspectives are presented. **Jalal Shayegan** and **Amir Badakhshan** present an Iranian perspective on problems generated by a rise in Caspian Sea level and ways to combat the adverse economic and social impacts of sea level rise. **Arif Mekhtiev** and **A. Gul** discuss environmental problems, explicitly pollution, from the perspective of the newly independent Republic of Azerbaijan, and propose some solutions to those problems. They also underscore the need for monitoring the environment in the Caspian region, much of which ended with the breakup of the Soviet Union. **Agadjan Babaev**, Director of Turkmenistan's Desert Institute, briefly presents the impacts of sea level rise along the Caspian coast of his country. He provides details on changes in the Kara-Bogaz-Gol Bay and the Cheleken Peninsula (now an island because of sea level rise). **Grigoriy Voropayev** discusses changes in the Caspian Sea basin on geological as well as historical time scales. He notes the importance of research on these time scales for oil exploration in the Caspian Sea and provides a schematic depicting, with varying degrees of probability, oil deposits under the Caspian Sea bed based on the inflow of ancient rivers. **Yuriy Chuikov** discusses the impacts of sea level rise and of pollution (from the perspective of a resource manager) at the mouth of the Volga River as it passes through Russia's Astrakhan Oblast. **Zalibek Zalibekov** provides the reader with information about the adverse impacts of sea level rise and pollution aspects of the Caspian from the perspective of the coastal Dagestan Republic in Russia.

Part III focuses on *Desertification* in the Republic of Kalmykia. (Kalmykia was to have been the location of this particular NATO ARW, but the location was shifted to Moscow for a variety of reasons.) Aside from the inundation of low-lying areas due to sea level rise, some of the countries bordering the Caspian are faced with desertification processes, or the creation of desert-like conditions where none had existed in the recent past. Desertification is the result of the interaction between natural processes and human activities, often inappropriate land-use practices related to cultivation and livestock overgrazing. The first paper by **Tatyana Saiko** and **Igor Zonn** is reprinted with the kind permission of the editors of *Geographical* magazine. According to these authors, Kalmykia is the site of Europe's largest desertified area. It presents an overview of the societal and environmental situation in a Russian republic that borders the Caspian. The second chapter in this section is the result of research by **I. G. Granberg and colleagues** on the origins of dust storms in Kalmykia's famous Black Lands. Much of the research over the past 60 years on soil and vegetation in the Black Lands has been carried out by famous Russian soil scientist Sergei Zonn, who in his 90th year prepared a background paper on desertification in Kalmykia for the workshop participants, a regional problem he has studied for much of his professional life.

Part IV contains chapters on suggested ways to cope with environmental change in general, as well as specific references to the Caspian. Ecological characterization provides an interesting approach to providing a baseline assessment of ecosystem(s) in the Caspian region. Once an ecological characterization has been developed and expanded to include a socioeconomic characterization of the forces affecting the region, it is possible to monitor ecological changes in the region. **John Kineman** and **Bradley Parks** present guidelines for undertaking such characterizations, with an example of its application to a coastal region in North America. **João Morais** presents an overview of biological and social global environmental change research in other parts of the world in the hope that similar research could be considered for the circum-Caspian region.

Victor Dukhovny and **Vadim Sokolov** present in brief a plan to resolve two water-related regional problems. They propose the transfer of water from the rising Caspian Sea to the declining Aral Sea. This is a renewed appeal to the seemingly dead issue of water diversions from either northward-flowing Siberian rivers or from the Volga River, an issue that many had thought had been put to rest in the mid-1980s. **Badakhshan** and **Shayegan** provide a chapter on concerns about the increasing risk of oil spills in the Caspian, as exploration and exploitation of oil deposits increase in the region.

Part V presents examples from other regions. **Münir Özturk**, **F. Özdemir** and **E. Yücel** discuss from a Turkish perspective several environmental problems in the Black Sea. If oil pipelines run from the Caspian to the Black Sea, there would be an increase in maritime oil transport and increased risk of accidental oil spills. They suggest sustainable uses of this unique marine ecosystem as well as lessons for managing the Caspian environment. **Sergei Preis** highlights many of the environmental problems faced by the littoral countries of the semi-enclosed Baltic Sea, a sea that has captured considerable international and regional attention during the past few decades. He suggests similarities in management of the Baltic and Caspian Seas. **Pavel Szerbin** addresses his concern about the contamination of the Caspian basin's rivers, sea, biota, and soils from radioactive materials and heavy metals. He notes that little, if any, monitoring of the basin has been undertaken since the Chernobyl accident and the breakup of the Soviet Union. He proposes future research opportunities in this area of concern.

The potential interest in regional or international organizations in the Caspian basin is presented in Part VI. Using Lake Leman (an international body of water shared by Switzerland and France) as an analogue, **Ilter Turan** attempts to draw parallels with the Caspian and discusses the problems and prospects for international cooperation among littoral as well as the larger number of basin states. **Abbasgholi Jahani** not only underscores the need for an effective regional organization for the Caspian but also provides some examples of the possible structure and function of such organizations, given the shared environmental resources and problems of the littoral states. **Gerhart Schneider** and **Mikiyasu Nakayama** present their respective views on the prospects for developing regional organizations in the region. Schneider identifies the needs for political cooperation in addressing the Caspian's environmental problems, while Nakayama raises the issue of whether there is a role for international organizations (such as the World Bank) in addressing Caspian environmental problems.

Acknowledgments

The editors would like to thank the participants of this ARW. Their participation was extremely valuable to the deliberations at the workshop. They were joined by other participants, such as Nina Novikova, Dr. Zaletayev, and Yuriy Starikov, among others. Special thanks also go to Yvette Zonn, Georgi Zonn, and Cendrik Zonn, who provided participants with the resources and transportation needed to make the workshop a success. On short notice (three days), they were able to establish a venue in Moscow for the workshop and to arrange the logistics for about 20 foreign participants. We also give our sincere thanks to Luis da Cunha, NATO Scientific Affairs Division, and to Isaac Skelton and Alla Rabinovich of the International Science Foundation. The editors appreciate the efforts of Justin Kitsutaka for his preparation of the graphic support for this volume. The lion's share of our gratitude goes to D. Jan Stewart, who arranged for the meeting's needs from the outset to the final preparation of the papers for this publication. Without her dedication to the Caspian project, this book would not have been produced in the twentieth century!

I. Introduction to the Caspian Sea Environment

THE REGIONALIZATION OF CLIMATE-RELATED ENVIRONMENTAL PROBLEMS

MICHAEL H. GLANTZ
Environmental and Societal Impacts Group
National Center for Atmospheric Research
Boulder, Colorado, USA

At a NATO Advanced Research Workshop (ARW) on Regional Organizations in the Context of Global Warming held in Paris in November 1992 [3], I prepared a short discussion paper on the regionalization of climate-related environmental issues. The reasons I wanted to talk about regionalization stemmed from a combination of factors: (a) my concern (along with that of a growing number of people) about the overfocus at that time on global climate change and the generation of numerous speculative scenarios for regional impacts of climate change, and (b) a book produced years ago by Joseph Nye, called *Peace in Parts* [9]. The idea at the time was to generate discussion about the possible utility for policymaking of the notion of the regionalization of environmental problems and responses to them.

Because of constraints of time and resources at the 1992 NATO ARW, the workshop focused on two major environmental issues of concern to many nations: (1) transboundary water resources, and (2) fish populations that crossed international borders at some stage of their life cycle. For the Caspian ARW, the notion of regionalization was applied to another climate-related environmental situation that involved more than one country: Caspian Sea level rise.

The Caspian Sea is really an inland terminal lake with no natural outlet to the world's oceans. Its level has fluctuated throughout time. The recent historical record goes back about 150 years. Before then, the record relies on geological, biological, and paleoclimatic proxy (indirect) information. In this period, some multiyear level changes have lasted several decades. The Caspian Sea began its most recent multidecade decline in the 1930s. The level dropped rather sharply at first, and slowly thereafter until 1977. In those decades, the sea had dropped about 3 meters. It began to increase in 1978.

Since 1978, the Caspian level has been on the rise, increasing about 2.5 meters in just under 20 years. This rise in sea level has caused the inundation of low-lying areas, destroying or threatening the viability of roads, rail lines, buildings, agricultural fields, villages, etc. Each of the states bordering the sea suffers to varying degrees from the rise in sea level. It is truly a regional commons problem. These states contribute to and/or

M.H. Glantz and I.S. Zonn (eds.),
Scientific, Environmental, and Political Issues in the Circum-Caspian Region, 3-10.
© 1997 *Kluwer Academic Publishers. Printed in the Netherlands.*

share, to varying degrees, the impacts of other environmental problems as well: pollution of the sea by oil and sewage, overfishing (of caviar-producing sturgeon), desertification. These states are focused on exploiting the Caspian's major natural resource: oil. Oil and sturgeon, sturgeon and oil. The Caspian is now known for the exploitation of these resources, both of which are nonrenewable.(At a recent CITES meeting [Convention on International Trade in Endangered Species], the fate of the Caspian Sea sturgeon was discussed for possible inclusion on the list of endangered species.[2].)

When the Soviet Union existed, there were only two littoral states: the USSR and Iran. Historical agreements between them and their predecessor governments provided rules that guided their interactions with regard to the sea. After the breakup of the Soviet Union in December 1991, the number of littoral states increased to five, as Turkmenistan, Kazakstan, and Azerbaijan gained independence from the Soviet Union. Whereas the Caspian had more or less been viewed as a Soviet sea, nowadays the sea belongs to no one; yet, in a way, it belongs to everyone. Without rules of conduct for the littoral states with regard to resource exploitation, the Caspian has become a classic common-property resource; that is, a resource exploited by many but under the jurisdiction of no one. Each coastal country can take the resources it wants (or pollute to the level it deems necessary) with little or no regard for the sustainability of the resources or the needs of the other countries with respect to those resources. Garrett Hardin [6] wrote about the "tragedy of the commons," using rangelands as his prime example. Over a decade earlier, Scott [12] wrote about the problems of a common property resource, referring to fisheries. This situation exists in the Caspian Sea with regard to sturgeon. Because of the impacts of the national competition among countries in the Caspian for sturgeon (and the highly valued caviar it produces) and high levels of illegal poaching, the sturgeon fishery is in a deep struggle for its survival.

Perhaps these shared problems of the nations of the Caspian region suggest that they could benefit from a regional environmental organization. Such an organization could generate a spirit of cooperation based on environmental protection of the Caspian Sea, and their shared (common) resources, while reducing the potential for conflict over the Sea's abundant, highly valued resources. They can put their energies into sharing the sea's resources—managing them efficiently, profitably, and sustainably.

An environmental organization for the Caspian could be composed of the five littoral states, or it could be constructed to include all of the states in the basin or even all of the provinces and autonomous republics bordering on the coast. Each state has an asset, a liability, or a problem to bring to a regionally focused table. For example, 80 percent of the river flow into the Caspian comes from Russia's Volga River and, as a result, decisions related to the Volga (volume, pollution, etc.) can have major impacts on both the Sea's level and its quality. For their part, Azerbaijan and Kazakstan have considerable known oil reserves and expertise. Some of the oil- and gas-rich states in the Caspian area must rely on other neighboring states in order to transport the oil from their wells to the world marketplace; and so on. The sturgeon spawn along the Turkmen coast and migrate to other parts of the sea.

In addition, there is still a need for a clarification of the legal status of the Caspian (i.e., Is it a lake or a sea?). Depending on the type of water body it is ultimately determined to be, different international legal standards would apply with regard to resource ownership, use, and responsibilities. It does, however, have characteristics of both a sea and a lake, having at one time been connected to the world ocean.

Regional organizations, with few exceptions, are considered to be ineffective. They are often (usually) dominated by one or two states, and collective action on resolving crucial issues is often subject to a veto. So, their levels of effectiveness are greatly constrained by national policies. During the 1950s and 1960s, regional organizations received considerable attention from the academic research community, as well as from national policymakers, as potential contributors to cooperation and peace around the globe. At that time, Joseph Nye wrote about the regionalization of world politics, capturing the "spirit of the times" in his book, *Peace in Parts* [9]. Citing other scholars, he noted that "the political life of the world is becoming less at the level of national states and more at the level of continents." It appears, however, that interest in regional organizations declined during the 1970s and 1980s, compared to that given to national interests and to the United Nations. However, the end of the Cold War, the emergence of numerous regional and local conflicts, and the bogging down of United Nations activities due to its expanded peace-keeping role enable us to reevaluate the potential contribution of regional environmental organizations to the well-being of the global community.

With the end of the Cold War and the decades-long rivalry and competition between the US and the USSR, there was considerable hope that the United Nations system would be able to operate as it was originally intended — as a global community of nations fostering cooperation and discouraging aggression. That hope was renewed as a result of the Persian Gulf War, when UN Security Council members apparently worked together (however reluctantly) to reverse the consequences of an Iraqi takeover of Kuwait. For a moment, such unity gave the impression that international cooperation (at least among the global powers as well as regional ones) would become the rule, and not the exception. However, that euphoria may have been premature, given the numerous military and political conflicts that have broken out in all parts of the world — even in southeastern Europe such as in the former Yugoslavia and Albania.

Perhaps, as we approach the beginning of a new century and a new millennium, leaders around the world will see this as a "window of opportunity" and, therefore, an incentive for an objective re-consideration of the potential contribution of regional organizations toward resource-related cooperation and conflict resolution. Given the resurgence of worldwide concern about "global" environmental issues and their regional causes or consequences, regional organizations with focused missions could provide an effective arena for discussing, resolving, or averting regional conflicts related to environmental change.

For its part, the scientific community working on global change issues is, as well, increasingly turning its attention toward the demands of political leaders for greatly improved, reliable, regional information about global climate change. They are less interested in global averages of temperature and precipitation changes than in national causes and effects of global warming. This scientific interest in regional details of climate change can be witnessed in the text of the most recent Intergovernmental Panel for Climate Change (IPCC) Climate Change Assessment [8].

It is necessary to note that natural or anthropogenic environmental problems can be viewed as global for different reasons. They could be global (and national) in cause (e.g., the onset of an Ice Age); they could be global in effect, although local in origin (e.g., ozone depletion, global warming, tropical deforestation, biodiversity loss); or they could be local in cause and effect BUT of worldwide interest (e.g., desertification, biodiversity loss). Why a problem has been defined as "global" will often determine whether as well as how it is dealt with by the community of states.

Although the consequences of global climate change will be (or, as some scientists now argue, are being) felt at the local and regional levels, global climate modelers are not yet in a position to provide societal and environmental impacts researchers with the regional detail needed for the development of reliable and credible scenarios of future climate at those spatial scales. Thus, national and subnational decisionmakers have become increasingly concerned about how a global climate change might manifest itself at the regional and local levels. The time may be right to talk about the "regionalization of environmental problems."

To date, most attention has focused on the causes of global warming. This has, in turn, sparked concern about the policy options countries might pursue to prevent, mitigate, or adapt to the consequences of global warming. After a long and arduous negotiating process, this concern generated the Framework Convention on Climate Change (FCCC) and led to the Conference of Parties (COP) to hammer out details of how countries might meet their obligations under the FCCC.

Attention by different researchers has also turned to developing various methodological approaches designed to gain a glimpse of the possible regional consequences of global warming of a few degrees Celsius. To date, most of the officially accepted scenarios about the potential regional effects of global warming have been based on the outputs of various general circulation models (GCMs). Because the precipitation scenarios produced by different GCMs disagree substantially for individual regions, their use for developing proactive policy prescriptions is risky. In addition, GCMs provide little in the way of reliable information about likely changes in the location, frequency, and intensity of extreme meteorological events.

Questions about the potential severity of impacts pertain not only to the physical climatic factors (including extreme events), but also to their relationship to societies possessing different capabilities for implementing proactive as well as reactive policy responses. These differences with regard to capabilities to respond raises the issue of climate-change-related "winners" and "losers." While it may be in the interest of some countries to say "no one loses with a climate change" (e.g., [1]), or that "climate change is not human-induced" (i.e., the Saudi Arabian position) or that there will only be losers if global warming occurs (e.g., [5], the rich industrialized countries), today we have a distribution of global climate in which some countries can be said to be winners and others, relatively speaking, losers. Thus, it is not difficult to argue that there are sure to be differential impacts on socioeconomic strata within society as well as among countries. A redistribution of regional climate regimes would likely produce a redistribution of climate-related winners and losers.

In addition to a reliance on GCMs for developing plausible scenarios of likely future regional climates, other approaches include the following:

- *Paleoclimatic reconstructions*: Past climatic periods, such as the Altithermal (4000 to 8000 years ago), may provide analogues to the expected climate of the next century. However, the forcing mechanisms that produced the atmospheric warming then were quite different from those proposed now. Paleoclimatic reconstructions must be used with caution.

- *"Warmest Arctic summers"*: This approach is based on the fact that GCM scenarios suggest that atmospheric temperatures would be notably higher in summer at high latitudes than at midlatitudes. Consequently, the global weather patterns of a set of previous years in the 20th century with unusually warm Arctic summers were identified and used to suggest possible climatic patterns in the midlatitudes under global warming conditions. Although an interesting approach, its reliability to regional scenario development has been scientifically questioned.

- *Historical 20th century analogues*: Sets of decades throughout this century have been warmer (1970 to present) or colder (1940-70) than average in certain regions. The responses of both ecological and societal systems during these periods can be examined to assess the regional climate sensitivity to decadal-scale climate fluctuations.

- *"Forecasting by analogy"*: Societal responses to recent extreme climate-related environmental changes (e.g., lake level rise in the 1980s in the Great Lakes, the Great Salt Lake, and, in the 1980s and 1990s, in the Caspian Sea) may be examined to identify weaknesses and strengths of present-day institutional responses. Such studies have, for example, suggested that rates of environmental change are often as important to decisionmakers as processes of change. There are benefits to be gained by learning about how well different societies cope with environmental change.

In sum, it is important to select with care those scenarios that may be proposed for use to define possible policy actions at the regional level. One must also realize that regional, counter-intuitive, climate-related changes cannot be ruled out. In fact, there have been several attempts to improve our understanding of potential surprises related to climate change.

Most likely, the impacts of a climate change on society will be directly felt as changes in the variability of climate, and, more specifically, as changes (or perceptions thereof) in the frequency, intensity, duration, and location of extreme meteorological events. Even with regard to adverse impacts of changes in the "global commons" (the atmosphere, the oceans, and even the tropical rainforests), the regional impacts of those changes will most likely manifest themselves in different ways and, perhaps, even in different directions. For example, the citrus-growing region of the US state of Florida witnessed at least five major freezes in the 1980s, the hottest decade on record. This example suggests that regional counter-intuitive surprises should be expected with a global warming of the atmosphere.

How then might regional organizations face an uncertain climatic future? Even in the absence of a CO_2-induced global warming, the global climate regime fluctuates on all time scales of interest to society. How might decisionmakers pursue sustainable development policies during interannual and decadal periods of climate-related environmental changes?

Whereas regional international organizations (functional as well as geographic) have often been relegated to marginal roles in the international political arena with regard to resources issues, this will likely change in the future. As resources and people dependent on them "migrate" on land and in the sea, the risks of regional conflicts, as well as the opportunities for regional cooperation, are likely to increase. This can be expected because, as climate regimes shift in response to global warming, so too will the location of some highly valued resources, sometimes across national boundaries. In addition, in locations such as the Caspian basin, the resources that are valued are in what could be considered a "regional commons." In the absence of clarification of the legal status of the Caspian, there is little coordination or cooperation, either in the protection of the Caspian resources or in their exploitation. Are environmental issues different in any way from other international issues? Are existing regional organizations prepared to deal with this "new" source of conflict?

Focusing on regional environmental problems and on the potential contributions of yet-to-be-formed regional organizations could provide useful insights into how well different types of regional organizations might fare in dealing with issues related to regional environmental changes such as Caspian Sea level rise, even in the absence of a global climate change. For example, 40 years ago, then Secretary-General of NATO,

Paul-Henri Spaak (1957-61) noted (with regard to the Anglo-Icelandic Cod Wars) the difficulties he had encountered in trying to resolve the resource-related conflict between the main protagonists (the UK and Iceland), both of which were key members of the political and military North Atlantic Alliance [13]. This particular conflict led to Icelandic threats to withdraw from use by NATO allies its strategically situated air base at Keflavik. It also led to hostile actions between two NATO members: Iceland and the UK [4].

A decade later, the report of the World Commission on Environment and Development (WCED) reinforced the view that it would be useful to reconsider the potential utility of regional organizations in coping with the impacts of climate-related environmental change. The report *Our Common Future*, also known as the Brundtland Report [14], drew attention to the Brundtland Commission's observation that, "policies formerly considered to be exclusive matters of 'national concern' now have an impact on ecological bases of other nations' development and survival." It recommended that "regional organizations in particular need to do more to integrate environment fully into their macroeconomics, trade, energy, and other sectoral programmes" (pp. 314-315), and noted that existing regional organizations must undertake considerations of environmental issues, calling on them to provide the appropriate level of interest, personnel, and financial support. In fact, in the late 1970s, the Brandt Commission report [11] stated that "it can no longer be argued that the protection of the environment is an obstacle to development. On the contrary, the case of the natural environment is an essential aspect of development" (p. 114).

The Brundtland Commission also briefly discussed regional organizations from the perspective of sustainable development, noting the strong need to better match international institutions with ecological problems. There is a growing awareness of the fuzziness of distinction between domestic and foreign policy (especially as one country, for example, can plunder its own natural resources, giving little regard to the global implications of its action). The Commission also suggested that intergovernmental regional organizations could gain considerable support from regional non-governmental organizations (NGOs), which bring different perspectives to roundtable discussions on environmental issues. The WCED briefly highlighted river basins as an example of international ecological systems in need of regional cooperation. The Caspian basin is a special example of a terminal lake.

Given that the frequency of such conflicts (or opportunities for cooperation) could increase in the face of global decadal-scale fluctuations of longer-term climate change and its regional consequences (e.g., sea level rise or changes in the productivity of fisheries) [7, 10], increased attention should be given to the role of regional organizations that might yet be established in resource-related regional conflict resolution. This multinational, multidisciplinary NATO ARW on the "Scientific, Environmental and Political Issues in the Circum-Caspian Region" was designed to encourage discussions among resource managers, scientists, representatives of regional organizations, political scientists, legal scholars, and geographers.

References

1. Budyko, M.I., 1996: Past changes in climate and societal adaptations. In: J.B. Smith et al. (eds.), *Adapting to Climate Change: Assessments and Issues*. New York: Springer-Verlag, 16-26.

2. CITES (Convention on International Trade in Endangered Species), 1996: The Multilateral Consultation on Problems of Sturgeon Species within the Framework of CITES. Proceedings, Conference held 21-22 November 1996 in Moscow, Russia. CITES Secretariat: State Committee of the Russian Federation for Environmental Protection. Contact Marco Pani, Secretariat, 15, Chemin des Anemones, Case postale 456, CH-1219, Chatelaine, Geneva, Switzerland.

3. Glantz, M.H. (ed.), 1994: *The Role of Regional Organizations in the Context of Climate Change*. Series I: Global Environmental Change, Vol. 14. Berlin: Springer-Verlag.

4. Glantz, M.H., 1992: Global warming impacts on living marine resources: Anglo-Icelandic Cod Wars as an analogy. In: M.H. Glantz (ed.), *Climate Variability, Climate Change and Fisheries*. Cambridge, UK: Cambridge University Press, 261-290.

5. Gore, A., 1992: *Earth in the Balance: Ecology and the Human Spirit*. New York: Houghton-Mifflin Co.

6. Hardin, G., 1968: The tragedy of the Commons. *Science*, **162**, 1243-1248.

7. Homer-Dixon, T.F., 1991: On the threshold: Environmental changes as causes of acute conflict. *International Security*, **16**, 2, 76-116.

8. Houghton, J.T., et al. (eds), 1996: *Climate Change 1995: The Science of Climate Change*. IPCC Working Group I Report. Cambridge, UK: Cambridge University Press.

9. Nye, J.S., 1971: *Peace in Parts: Integration and Conflict in Regional Organization*. Boston, MA: Little, Brown & Co. Reprinted in 1987 by University Press of America.

10. Perelet, R., 1988: Ecology and diplomacy. *International Affairs* (November), 86-94.

11. Report of the Independent Commission on International Development Issues, 1980: *North-South: A Programme for Survival*. Cambridge, MA: MIT Press, p. 114.

12. Scott, A.D., 1955: The fishery: The objective of sole ownership. *Journal of Political Economy*, April, 116-124.

13. Spaak, P.-H., 1971: *The Continuing Battle: Memoirs of a European 1936-1966*. London, UK: Weidenfield and Nicholson.

14. World Commission on Environment and Development, 1987: *Our Common Future*. New York: Oxford University Press, 311-315.

CLOSED AREAS AND THE CASE OF THE CASPIAN SEA BASIN

GENADY N. GOLUBEV
Faculty of Geography
Moscow State University
Moscow, Russia

In a considerable part of the world rivers do not flow to the ocean but, instead, they flow into inland drainage basins. Such basins have unique natural, socio-economic and geopolitical features. Because of the interactions between natural and socio-economic factors, many of the enclosed seas (or lakes) are subjected to environmental tension and in some cases environmental disaster. It is amazing that in spite of the large spatial extent of these enclosed basins and their importance in the Earth System, there are no interdisciplinary publications on the subject which sum up their main features. A proposal to organize an international project to undertake such a study deserves serious consideration. This chapter is an attempt to take a first step in this direction and to identify the main features of the enclosed drainage basins of the world.

Since the Caspian Sea with its basin is the largest enclosed area in the world, there are reasons to believe that an analysis of unique and common features of these enclosed areas might be useful in addressing Caspian problems.

Often, enclosed drainage areas and areas without any river flow are considered to be synonymous. They are not. There are three categories of land from the point of view of the main characteristics of runoff: *exorheic* areas provide runoff to the ocean, *endoreic* areas provide runoff within enclosed areas, and *arheic* areas are those which do not provide any runoff at all. Strictly speaking, the last two are areas where runoff rarely occurs. It is impossible to draw a reliable line dividing endoreic and arheic territories on a map or even in the field, because of the weakness or virtual absence of fluvial processes and, hence, the weakness of the fluvial geomorphologic features which is the basis on which the dividing line is drawn. That is why on the maps of continents endoreic and arheic areas are combined. Still, the drawing of the dividing line has a number of methodological difficulties which deserve special discussion in the future. The area by continent of endoreic and arheic territories is as follows:

M.H. Glantz and I.S. Zonn (eds.),
Scientific, Environmental, and Political Issues in the Circum-Caspian Region, 11-16.
© 1997 *Kluwer Academic Publishers. Printed in the Netherlands.*

TABLE 1. Endoreic and Arheic Territories by Continent

Continent	Area of continent (million km^2)	Enclosed areas (million km^2)	Enclosed areas % of continent
Eurasia	54.0	15.9	29
Africa	30.1	12.9	43
Australia	7.6	4.2	55
South America	17.8	1.6	9
North America	24.4	0.9	4
The world in total*	**133.9**	**35.5**	**27**

* Excluding Antarctica, Greenland and the islands of the Arctic.

The total area of the endoreic and arheic territories in the world is 35.5 million km^2, or about 27% of the ice-free land surface. The State Hydrological Institute [1] gives a figure of 29.8 million km^2, and Lvovich [2] determined it to be 32.1 km^2.

The Caspian Sea basin is endoreic, but inside there are arheic spots within it. The Caspian, including both the Sea and the basin, is the largest enclosed (endoreic) territory in the world with an area of 3.1 million km^2 or about 10% of all enclosed territories in the world.

The main common features of enclosed areas are the following:

a) A bowl with a relatively flat bottom can be taken as a morphologic (physical) model of an enclosed area containing a lake or a swamp. Because of its morphology, an endoreic area is an integrated natural and social system where the main system-forming relations are the runoff processes. This means that within the system in question most fluxes of matters are predominantly in the downward direction. The natural processes and their fluctuations in the zones of runoff formation exert considerable influence on the natural and socioeconomic state of the lower zones situated closer to the bottom of the "bowl" — the belt of runoff dissipation. Therefore, the physical links downward in an enclosed system have important economic and geopolitical consequences.

b) Both endoreic and arheic areas are usually situated in arid lands where evapotranspiration is considerable higher than precipitation. In such areas agriculture is possible only with irrigation. The development of irrigation leads to changes in the water balance and in the water regime. It is the driving force in the human-induced transformation of the landscape. Usually, irrigation degrades the system's environment.

c) Under natural conditions, the water level of a terminal lake is a good indicator of the degree of water availability in its entire basin. The inflow to the lake from various locations within the basin is the most important component of the water balance, usually determining the natural fluctuations of water level. The local water cycle (the one within the enclosed basin) does not normally play a significant role in the basin's water regime, even for large basins. Natural fluctuations in runoff are basically random in nature. Water resource management in the basin usually involves a trend toward growing water withdrawal and consumption. Thus, ultimately the water level of a terminal lake is the result of the combination of natural and anthropogenic factors.

d) Significant and relatively sharp variations of water level are characteristic of all closed lakes and seas of the world. Perhaps, it is the most significant feature of an endoreic system. For the Caspian Sea, a typical variation of the water level is about 1.5 m in a 10-year period or 10 m in a 1,000-year period. Economic activities at the shoreline of the lake usually cannot cope with such sharp oscillations without considerable loss. Therefore, development of a long-term forecast (e.g., 10-20 years ahead) of the water level of closed lakes is a very important problem worldwide. Keeping in mind the limited success in forecasting terminal lake water-level, the formulation of a strategy for economic development and activity under conditions of environmental uncertainty (e.g., lake-level variations) is a high priority.

e) There are general tendencies in the world's endoreic areas: the increase of human-induced pressure on natural systems, and increasing aridization as a naturally occurring background condition. Changes in the total volume of water resources can serve as an indicator of human pressures as well as of climatic changes. Changes in the total amount of water resources can provide useful information on the availability of water resources as well as on the degree of aridization.

f) Water is the most precious resource in endoreic and arheic areas. Aside from its quantity, the key problem in these areas is the optimal use of water resources as a part of a sustainable development strategy. In such areas the per capita volume of water resources is declining, while water demand continues to grow. The major conflict in an enclosed basin area is that over water use among different, competing administrative, political and cultural entities within the enclosed basin. These kinds of conflict will likely become ever more intense, and could lead to very serious political and, as some suggest, armed disagreements.

g) Any significant human action within an enclosed area, such as the plowing up of new lands, the construction of water reservoirs, the construction of irrigation and drainage networks, population growth, and so forth, cause a large spatial distribution of chemical substances and produces soil erosion as well. In sum, these human activities often lead to a major widespread transformation of landscapes.

h) The strategy to control the environmental state of a terminal lake cannot be successful without developing a strategy for its entire basin.

i) Natural variations in the surface area and level of terminal lakes depend on hydrometeorological and geological (e.g., tectonic) factors. Moreover, the longer the period under consideration, the more important becomes the influence of geological factors.

j) Changes in the hydrometeorological conditions of the endoreic and arheic areas during the historical period and the recent geological era used to lead to considerable changes in the levels of terminal lakes and, as a consequence, to the drastic changes in regional geographical conditions. The literature is quite rich with data on changes in water level of such closed lakes as the Caspian Sea, Aral Sea, the Great Salt Lake, Lake Chad, Lake Eyre, and the like. There is also information on the "opening" of some enclosed areas during certain, usually pluvial periods, suggesting that Lake Chad had at one time been connected to the Mediterranean, and that the Amudarya once flowed to the Caspian by way of the Uzboi channel (a series of depressions across the Karakum Desert). Most probably, some arheic areas have turned into endoreic or exorheic ones, depending in part on their topographical features.

k) Due to their peculiarities, enclosed areas are to a great extent (but not completely!) decoupled from the global cycles of matter, so that about one fourth of the land surface functions autonomously. Taking this into account is important for understanding the working of the Earth System.

l) The structure of any enclosed area inevitably leads to potential for a conflict situation with regard to the use of water resources: the water users situated upstream are not primarily interested in the quantity or the quality of water that they allow to flow downstream; nor are they concerned about the environmental consequences downstream of their effluents. A sustainable (e.g., long term) solution of this potential conflict is possible, only if all participants consider the system holistically and, on this basis, carry on concerted actions.

m) If an enclosed (endoreic) area is shared by a few states, only serious international cooperation in managing this system can assure its sustainablility.

The features summarized above are common to all enclosed areas in the world, including the Caspian Sea. At the same time, the Caspian has a few unique features.

The distinctive feature of the Caspian Sea is that it is the only large endoreic area in the world which has the predominant part of its watershed situated away from the arid regions and in areas where precipitation is approximately equal to or exceeds evapotranspiration. From natural science and geopolitical points of view, it is a very important characteristic.

The Volga River provides about 80% of the water inflow to the Caspian Sea, and fluctuations of the Volga's runoff are the main factors in variations of the sea's level. The principal part of Russia's population and economic activity is situated in the Volga basin. For centuries the Volga has been the center of mass for the Russian nation. Even the capital of Russia is within the Caspian basin. Thus, the economic and political actions and interests of Russia are the most important factors in controlling the Caspian water regime, including its water level. The geopolitical role of Russia in the Caspian region is much greater than the 10% of the Caspian coastline which formally belongs to Russia would suggest.

Because of the break-up of the former Soviet Union in December 1991, the Caspian coastline is now shared by five independent states and the basin is shared by eight states. Conflicts will inevitably emerge. Their peaceful resolution will be possible only through international cooperation. Such potential conflicts include those related to the use of the most precious natural resources of Caspian, namely fish and oil. It is desirable as well to analyze carefully the experiences gained from cooperation and from the problem-solving of conflicts around the world where there are other international terminal lakes, such as the Aral Sea, Lake Chad, the Dead Sea, Lake Turkana, and Lake Titicaca.

The Caspian Sea is so large that serious international law problems have arisen: should the Caspian be treated as a sea or as a lake. From both an environmental and a sustainable development point of view, the lake and the enormous watershed system in which it is embedded should be treated as one entity.

The break-up of the Soviet Union and the recent increase in the number of states linked directly to the Caspian Sea have led to a number of environmental problems including the loss of the environmental monitoring system for the sea. As a result, the condition of the lake is worsening but at what rate and to what degree we no longer know. A Caspian Sea environmental program that includes the renewal of a holistic monitoring system can be an effective step toward the stabilization and improvement of the Caspian environment.

Development of environmental indicators and indices of sustainable development is currently a very popular topic. Similar region-specific activities focused on the Caspian Sea region should be considered. Indices for the state of its environment will be a very useful tool for periodically measuring the effectiveness of international cooperation at the regional level.

In sum, a review of problems either common to all enclosed basins of the world, or specific to the Caspian Sea, provides a basis for suggesting a few priority issues in need of attention.

a) *A strategy to develop an international cooperation plan for the Caspian* (the concept; actions to increase awareness and trust; the tools of environmental management; indicators of the state of the Sea, etc.)

b) *The level of the Caspian Sea* (long-term forecasts; strategic plan for economic activity in the coastal zone under conditions of uncertainty).

c) *Exploitation of management of key resources of the sea and the basin*: fish, oil, water (political and environmental issues, tradeoffs).

References

1. Anon., 1974: *The World Water Balance and Water Resources of the Earth*. Leningrad: Gidrometeoizdat. 638 pp. (In Russian.)

2. Lvovitch, M.I., 1986: *Water and Life*. Moscow, Russia: Mysl. 255 pp. (In Russian.)

OVERVIEW: HISTORY AND CAUSES OF CASPIAN SEA LEVEL CHANGE

GEORGE S. GOLITSYN
Institute of Atmospheric Physics
Russian Academy of Science
Moscow, Russia

The Caspian Sea is the world's largest inland water body with an area of almost 400,000 km^2, corresponding to the combined areas of Germany and Netherlands. Its watershed encompasses 3 million km^2. Around 80% of the river flow to the sea comes from the Volga River. The coastline is close to 7000 km in length, only 10% of which belongs to the Russian Federation.

Geological evidence attests to the fact that the level and the area of the Sea are always varying [2, 3]. Better known is the history of the sea for the last 10,000 years [4]. During this time, the level was changing between −34 and −20 m (like several closed lakes, the level is below the world ocean level); 5 times it was above −27 m (Fig. 1). This has been reconstructed using paleoinformation.

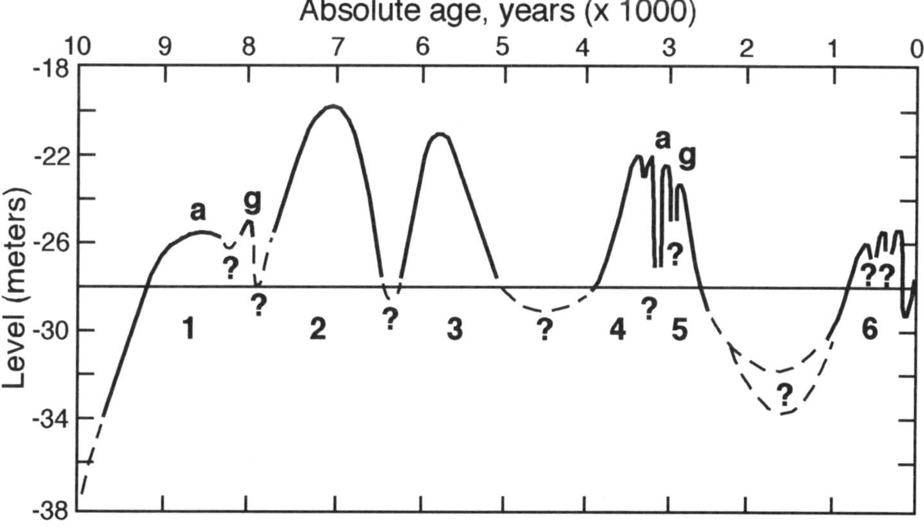

Fig. 1. Caspian Sea level for last 10,000 years

M.H. Glantz and I.S. Zonn (eds.),
Scientific, Environmental, and Political Issues in the Circum-Caspian Region, 17-25.
© 1997 *Kluwer Academic Publishers. Printed in the Netherlands.*

Instrumental measurements of the sea level began in 1837 first at Baku; shortly afterward Azerbaijan became a part of the Russian Empire. In the Soviet Union nearly 20 posts supplied the measurements. Figure 2 presents the composite record of the level for the 1837-1995 period. During the first 90 years, the level was within the limits of -25.7 ± 0.5 m. In the 1930s a sharp drop was observed (-1.7 m) in one decade. After 1940, a further drop with fluctuations occurred, reaching its lowest point (-29.0 m) in 1977. Since then, there has been a continuous rise (up to -26.6 m in 1994). In 1995, despite below normal precipitation in the Volga watershed, the mean annual level rose by another 13 cm. [NB: One should keep in mind that there are seasonal fluctuations in the level with the highest values in June and the lowest values in December, differing by about 40 cm, occasionally more].

After World War II, water from the Volga and other rivers was increasingly used for irrigation and other purposes. In the 1980s, according to the data of the State Hydrological Institute in St. Petersburg, the volume of the "used" water reached 43 km^3 annually which is equivalent to 11 cm for the sea level. The dashed line in Fig. 2 shows what the level would have been, if the water had not been diverted to other uses. Now, the level is 2.5 m higher than in the late 1970s, surpassing the biggest changes in the 19th century.

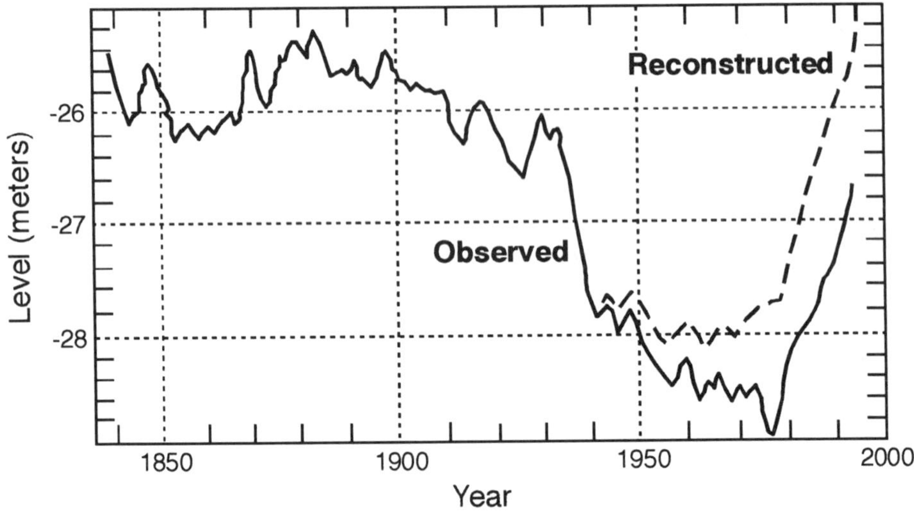

Fig. 2. Effect of irrigation on Caspian Sea level, 1800-1995

In the 1940-1980 period, several hydropower stations with reservoirs were constructed along the Volga River. Filling them up influenced the river's flow into the sea. Over the years, the volumes in the reservoirs have more or less been stabilized and their presence no longer plays a major role for the sea's level. An exception might be 1996, because the water volume during 1995 had been used up, to a large extent because of the low level of precipitation in 1995. However, the large snowfall during the past winter (1995-96) should ameliorate that condition. Thus, relatively small changes are expected in 1996.

The problem of calculating the sea level, using the data of several gaging stations, is not a trivial one. Wind surges, the effect of "inverse barometer", and a number of other factors need to be taken into account. There are vertical movements along the coastline with rates up to 1 cm/year. Precise knowledge of those movements is necessary for any long-term monitoring of sea level. The most accurate and economical way to determine the positions of the stations involves the use of the Global Positioning System (GPS).

When the Caspian Sea level was decreasing, there was a major concern about it. In the 1970s and up to 1986, major plans were being prepared to divert to the Volga part of the water of rivers that flowed naturally into the Arctic Ocean. During the first stage, 19 km^3 were to have been taken from the Pechora River, the North Dvina River, etc., through a system of canals. Heated debate raged in the 1980s about their economic value and the environmental threats associated with such diversions. In July 1986 plans were discussed at a USSR government meeting and, despite the fact that a vast majority of the invited experts were for a small start, the Chairman of the Council of Ministers, N.I. Ryzhkov, summing up the discussion, said that there were no funds for the work and proposed more research. This was at a time when the sea had already risen for eight years in a row and was a meter higher than in 1977 [4].

In the late 1970s, a plan was developed to dam the Bay of Kara-Bogaz-Gol. The bay was connected by a narrow passage to the Caspian Sea, drawing from it at the time about 5 km^3 of sea water per year. When the sea was at higher levels earlier in the century it was drawing up to 30 km^3. In 1980 the dam was quickly built. In April 1992 one of the first major decisions by the newly independent Turkmenistan Republic was to destroy the dam. During its twelve years of existence, the dam added about 0.5 m to the rising sea level. At the opening of the dam, already subjected to the impacts of high sea level, large volumes of water washed away material from the passage's bottom. Because of this, the bay is now a larger sink for sea water, consuming about 40 km^3 annually. This corresponds to about 10 cm of sea level. The continued rise of the sea would also lead to the filling up of natural (topographic) depressions in Kazakstan. These filled depressions would evaporate an additional 10-15 km^3 annually.

In order to understand the causes of sea level change, one should first know the variability of its water-balance components. The annual mean river flow is usually measured with an accuracy of a few percent. Precipitation had been measured at about 15 stations (in the time of the former Soviet Union) in the coastal zone and at 3-4 islands, together with wind, air temperature and humidity and, in some places, water temperature. These data together with the irregular collection of data from ships were used by Panin [13] to estimate evaporation from the sea's surface.

Together with flow data, the data for the 1930-86 period have been used by Golitsyn and Panin [7] to calculate sea level changes and these estimates were compared with the actual level changes. The correlation coefficient between calculated and observed level values was 0.87. Cumulative changes for the periods 1930-1939 and 1977-86 were found to be even closer to reality. The calculated values were 175 cm compared to the actual 170 cm for the first decade and 103 cm compared to 117 cm for the second decade. This correspondence shows the random nature of the errors in calculating the level changes. It also shows that level changes in a first approximation can be evaluated from the sea-water balance, without taking into account hypotheses on changes of the bottom topography due to recent tectonic activity [11] or due to sucking-in or compression-out of huge amounts of water by the sea-bottom sediments [17].

As one could expect, the causes of the fall in sea level in the 1930s and rise since 1977 were quite the opposite. In the first case it was the result of a decreased Volga River flow and increased evaporation from the sea's surface; now it is an increase in flow and a decrease in evaporation. As was shown by Golitsyn et al. [8] and Panin et al. [14], the decrease of evaporation is related to negative trends of the wind velocity over most of the sea and its watershed. The negative trends of evaporation have also been found by Peterson et al. [15] for 1950-90 in the European part of the former Soviet Union, Siberia, and eastern and western states in the USA. These trends have been associated with a decrease in diurnal temperature range changes and an increase in lower level cloudiness. Possibly, both causes are operating.

The Problem of Caspian Sea Level Forecasts

Since 1985, the Hydrometeorological Center of Russia has prepared forecasts of Caspian Sea level for a year in advance. The forecasts are issued in April, using the amount of snow and soil moisture in the Volga watershed as predictors. Their accuracy has been within 5 cm; for the 1995 forecast they projected 14 cm, while the real rise in sea level was 13 cm. There are attempts to lengthen the forecast period up to 1.5-2 years [12] based on climate statistics.

There have been attempts for much longer-lead forecasts based on solar activity, configuration of the planets, etc.; a review of them was undertaken by Ratkovich [16]. One can appreciate the futility of such attempts, recalling that for almost a century since the first use of instrumental measurements since 1837, the level had not exceeded 25.7 ± 0.5 m.

A scenario approach was used by Budyko et al. [1], applying paleoclimate analogies of the Holocene Optimum to the 1990s (the mean global temperature then was 1°C higher than in the mid-20th century), the Eemian (2°C higher) for the 2020s and the Pliocene (3° to 4°C higher) for the 2050s. They found for the 1990s a decrease by 0.3 m (compared to the 1980s), a rise by 1.5 m in the 2020s, and a rise of 5.4 m for the mid-21st century. One could argue about the numbers, but they can serve as an important warning.

A reliable forecast is needed now for local and central government authorities. See the dashed line in Fig. 1. If water had not been withdrawn from rivers, the increase in Caspian sea level would have started 15 years earlier. One can use the sea-water balance equation, using mean values for the last 20-30 years. The equation can be integrated analytically [10, 16], taking into account the linear relationship between the surface area of the sea and its level and the Kara-Bogaz-Gol Bay acting as a sink. Thus Kazansky estimated that in 2005 the level could reach −26.0 m and in 2029 it could be −25.5 m, close to the asymptotic level of −25.3 m e.g., similar to the values of the components for the last couple of decades.

An important question is "Will the components be stationary?". The hope is that a reliable answer can be obtained using large climate models. Their ability to reliably predict regional climate changes under conditions of increasing concentrations of greenhouse gases is still in its infancy. The climate modeling community around the world is striving for this goal. The Caspian case could provide the biggest regional climate challenge to address, because of the large area of the sea and its drainage basin and because of the major adverse impacts caused by the level rise.

Global climate models are now involved in the Atmospheric Model Intercomparison Project (AMIP). More than 30 models from various countries are participating in AMIP. For the period 1979-1988, they are using observed values of monthly mean sea surface temperatures and sea ice, observing how the atmosphere and land surface behaved during this period. Temperatures, clouds, precipitation, evaporation and river flow values in the models are compared with observations. The models of Russia's Main Geophysical Observatory and the Institute of Numerical Mathematics are involved in AMIP.

AMIP has a large diagnostics sub-program. Its Project # 24 is focused on "The Caspian Climate," of which this author is the leader. Considerable work is also performed at the Main Geophysical Observatory in St. Petersburg under the guidance of Prof. V.P. Meleshko [9]. About half of the model results have been diagnosed. Analysis shows that all models reproduce the rise, but only the models with a high spatial resolution reproduce the rise more or less realistically. This work continues with the assistance of the Max Planck Institute for Meteorology in Hamburg, Germany and the Hadley Center of the UK Meteorology Office. When the diagnosis of model runs has been completed, there are plans to analyze the results of century-long integrations under various greenhouse scenarios. So, a picture of how the sea might behave in the next century is still a long time away, relatively speaking. It seems that a clearer picture might emerge only from an integration of the whole ensemble of scenarios from the various climate models.

Major Problems Related to Caspian Sea Level Rise Impacts

All the impacts, especially those on human activities, stem from the fact that people have forgotten the laws that had been enforced for centuries in the ancient states along the coast of the sea: It was forbidden under the threat of death to construct buildings within a certain distance from the sea. People then and now know of its changing nature. After World War II, people started to move in (encroach) on the receding shoreline, setting themselves up for future sea-level-related problems.

The estimates of damage caused to all five states around the sea were presented in May 1994 at a UNEP Workshop on the Caspian in Moscow. They ranged between $30 to $50 billion US. Because the damage is a strong non-linear function of the sea's level, this number since that time could have doubled. Beaches have gone, port facilities to a large extent have been severely damaged, roads and buildings destroyed, basements full of water, groundwater levels have risen, etc. In Baku, the capital of Azerbaijan, the central square with government buildings facing the sea is now protected by a concrete wall. The protective construction is almost always in the process of destruction by sea waves during storms. An especially bad situation is on the low-lying coast of the northern Caspian. The wind surges are up to 3-4 m occasionally and are often up to 2 m. During these episodes, the sea water is driven inland up to several tens of kilometers. In March 1995 the Kalmykian town of Lagan was flooded for several days during such an episode.

Thus, the protection of endangered towns and cities is the highest priority, especially in its technical and engineering aspects. The smaller villages are likely to be relocated; by May 1994 about 30,000 people had already been moved away. Under the poor economic conditions of the former Soviet Union countries, these people are inadequately reimbursed, causing additional social tensions to those already generated by an absence of jobs and housing.

In early 1996 the Russian Government decided to allocate 483 billion Rubles (about $100 million US at that time) for urgent protective measures. What portion of this sum can be obtained in reality and how it can be used remains a big question these days in Russia. This was supposed to be a part of much larger Russian Federal Program for 1996-2000, the concept of which had been developed by the Interagency Scientific Council for Caspian Sea Problems. The concept and the program also proposed to do research, monitoring, etc. However, the funds allocated for 1996 covered only about 20% of what was needed.

The second most important problem is how the coastal economics should develop and what might be the people's employment. One thing is clear: the local authorities should remember the medieval warning: avoid construction near the sea's edge (at the level below, say, −20 m).

A host of environmental problems is appearing and sea water pollution is a major one. The soils that are inundated are a major source of pollution from fertilizers, sewage, etc. Reed plants occupy thousands of square km in the shallow northern parts of the sea. When they are inundated they rot, causing organic pollution, a deficit of oxygen, and hence, the death of plankton, fish and other organisms and are a threat to ecosystems. The damage to the sewage systems pose a severe threat to human health, e.g., cholera and the plague. The increase in the area encompassed by newly formed wetlands leads to a substantial increase in mosquito infestation and in malaria.

The Caspian Sea region is an area with vast oil and gas resources. There are plans to increase production including open-sea exploitation. This is a major potential threat to the health of the sea's ecosystem, if exploration and exploitation are not properly undertaken.

A proper legal framework for the Caspian Sea has yet to be developed (see Vinogradov, this volume). This has caused friction among littoral Caspian countries in all aspects of economic activities. The monitoring system of meteorological and hydrological parameters, including the sea level, has in the past few years deteriorated badly.

On June 20-23, 1995, an International Conference on "The Caspian Region: Economics, Ecology, Mineral Resources" was held in Moscow, and was attended by about 150 scientists and businessmen from Russia, Kazakstan, Iran, Germany, Italy, and the USA. More than one hundred papers were presented at the conference. Special discussions were held to identify the most important aspects of the Caspian problem.

The text of the "Call" is as follows:

CALL of the International Scientific Conference on the "Caspian Region: Economics, Ecology, Mineral Resources" to the Governments of the Caspian states, international organizations, and scientists and specialists interested in the national use and protection of the natural resources of the region.

Participants of the Conference listened and discussed papers that had been presented and came to an understanding of the following:

- The Caspian Sea is a unique natural formation, unified in the sense of ecological and natural resources, requiring joint management aimed at rational use and protection of its resources.

- Integrated studies are needed of the Caspian region based on combined international efforts, first, concerning the Caspian states.

- The solution of common problems of the region needs the priority attention of the authorities of the coastal states and requires a corresponding legal framework.

- To organize integrated regional studies, it is necessary to strengthen the existing monitoring systems, including geodynamic monitoring, to build on their base a unified system of monitoring of environmental and biological resources, to provide for the preservation and continuation of the observational series and the free exchange of data and results of scientific studies among specialists and scientific organizations of the Caspian States. The experience of ecological work in the Black Sea region shows a necessity to have a united coordination center for this.

- In the interest of minimizing the total damage and increasing the ecological-economic efficiency of natural resource use in the Region, it is necessary to organize the joint international expertise of the projects related to the use of the region's natural resources.

- Implementation of coastal protection measures in various countries requires their mutual adjustment and exchange of experience, the joint evaluation of their effectiveness and their influence on the environment.

- It is necessary to work out together the concept of the sustainable development of the Caspian region, based on the results of fundamental and applied studies of Caspian Sea level forecasts, climatology, geology, geodynamics, natural resources, environment, ecosystems, economics, and legal aspects. The concept should determine the principal directions and foundation of the cooperative solution of the problems of natural resource use and protection, adaptational use of coastal areas, and priority directions for research in the near and long term.

- To organize the next international conference in the next few years and to ask the Interagency Scientific Council on Caspian Problems (e.g., the Russian Ministry of Science and Technological Policy, the Russian Academy of Science, and the State Committee for Water use and Resources of Russia) to make technical and logistical preparations for a 1998 Conference.

References

1. Budyko M.I., V.A. Efimova, and V.V. Lobanov, 1988: The future level of the Caspian Sea. *Meteorology and Hydrology*, **5**, 86-94.

2. Fedorov, P.V., 1994: Some problems of geological history of the Caspian Sea. *Stratigraphy and Geological Correlation*, **2**(2), 71-79. (In Russian)

3. Fedorov, P.V., 1995: On The Causes of Periodical Changes of the Caspian Sea level. *Herald (Vestnik) of Russian Academy of Sciences*, **9**. (In Russian)

4. Golitsyn, G.S., 1995a: The rise of the Caspian Sea level as a problem of diagnosis and prognosis of the regional climate change. *Izvestia - Atmospheric and Oceanic Physics*, **31**(3), 385-391.

5. Golitsyn, G.S., 1995b: Caspian Sea rising. *Novy Mir* (New World), **7**, 200-207.

6. Golitsyn, G.S., and G.N. Panin, 1989a: On the water balance and contemporary changes of the Caspian Sea level. *Meteorology and Hydrology*, **1**, 57-64.

7. Golitsyn, G.S., and G.N. Panin, 1989b: Once more about changes of the Caspian Sea level. *Herald (Vestnik) of USSR Academy of Sciences*, **9**, 59-63. (In Russian)

8. Golitsyn, G.S., A.V. Dzyuba, A.G. Osipov, and G.N. Panin, 1990: Regional climate changes and their evidence in the present Caspian Sea level rise. *Proceedings (Doklady), USSR Academy of Sciences*, **313**(5), 1224-1227. (In Russian)

9. Golitsyn, G.S., V.P. Meleshko, A.V. Meshcherskaya, I.I. Mokhov, T.V. Pavloa, V.A. Galin, and A.O. Senatorsky, 1996: *Proceedings of the AMIP Conference*, Monterey, CA, held in May 1995. (in press)

10. Kazansky, A.B., 1994: A possible approach to the problem of the Caspian Sea level rise. *Proceedings (Doklady), Russian Academy of Sciences*, **338**(4), [PAGE NOS] (In Russian)

11. Lilienberg, D.A., 1993: Tendencies of contemporary endodynamics of the Caspian and its level changes. *Proceedings (Doklady), Russian Academy of Sciences*, **331**(6), 745-750. (In Russian)

12. Meshcherskaya, A.V., and N.A. Alexandrova, 1993: Forecast of the Caspian Sea level using meteorological data. *Meteorology and Hydrology*, **5**, 86-94.

13. Panin, G.N., 1987: *Evaporation and Heat Exchange of the Caspian Sea*. Moscow, Russia: Nauka Publishing House. 86 pp. (In Russian)

14. Panin, G.N., A.V. Dzyuba, and A.G. Osipov, 1991: On possible causes of evaporation changes during last decades in the Caspian Sea region. *Water Resources, 3*, 5-17.

15. Peterson, T.C., V.S. Golubev, and P.Ya. Graisman, 1995: Evaporation losing its strength. *Nature*, **377**, 687-688.

16. Ratkovich, D.Ya., 1993: *Hydrological Bases of Water Supply*. Moscow, Russia: Water Problems Institute. 428 pp. (In Russian)

17. Shilo, N.A., and M.I. Krivoshei, 1989: Relationship between oscillations of the Caspian Sea level and stresses in the Earth's crust. *Herald (Vestnik) of USSR Academy of Sciences*, **6**, 83-90.

ASSESSMENT OF THE STATE OF THE CASPIAN SEA

IGOR S. ZONN
Russian National Committee for UNEP (UNEPCOM)
Moscow, Russia

The Caspian Sea is in a closed basin in the inland part of Eurasia. The sea's level is below that of the World Ocean. The sea basin stretches from north to south almost 1,200 km and its width varies between 200 and 450 km. The total length of the coastline is 7,000 km, and its water surface area is about 390,600 km^2 (as of January 1993). Water salinity in the northern part is 3-6‰, and in the middle and southern parts it is 12‰. Territorially, the sea is located within the boundaries of Russia, Kazakstan, Turkmenistan, Azerbaijan and the Islamic Republic of Iran (Fig. 1).

Sometimes the Caspian Sea is called "A Hard Currency Sea," because of the availability of two kinds of "black gold" — oil and caviar. Oil reserves in the region, including the Caspian shelf, contain no less than 18 billion metric tons and 6 billion m^3 of gas. During the 1980s, Soviet enterprises produced up to 2,500 metric tons per year of black caviar, and the Iranians produced 250 metric tons per year. However, according to the assessments, in 1995 the Russian enterprises produced only 70-90 metric tons of sturgeon caviar, and Iran only 200 metric tons [20].

The Caspian sturgeon makes up 90% of the world reserves of white sturgeon, sturgeon and starred sturgeon. Currently, the reserves of sturgeon of the states of the former USSR in the Caspian basin are in a critical condition. The habitats of Persian sturgeon are near the Iranian shores. The reduction in sturgeon became noticeable after the disintegration of the USSR, when the centralized system of control of sturgeon catches was ended. This resulted in the rapid emergence of poaching. Poachers hunt sturgeon only for its caviar. Today, they catch sturgeon directly in the open sea. However, in the early 1960s, prohibition was introduced by the former USSR against catching sturgeon in the open sea. Since that time, catching sturgeon has been carried out in the river deltas. It should be noted that sturgeons reproduce very slowly: the fish do not spawn for the first time until they reach the age of 20-25 years. In 1990, the permissible catch of sturgeon in the USSR was set at 13,500 metric tons. In 1996 it was only 1,200 metric tons [20].

The surface water inflow into the sea is formed by the flow of the Volga, Ural, Terek, Sulak, Samur, Kura, small Caucasian rivers, and the rivers of the Iranian coast. The Volga River, the basin of which makes up nearly 40% of the territory of

M.H. Glantz and I.S. Zonn (eds.),
Scientific, Environmental, and Political Issues in the Circum-Caspian Region, 27-39.
© 1997 *Kluwer Academic Publishers. Printed in the Netherlands.*

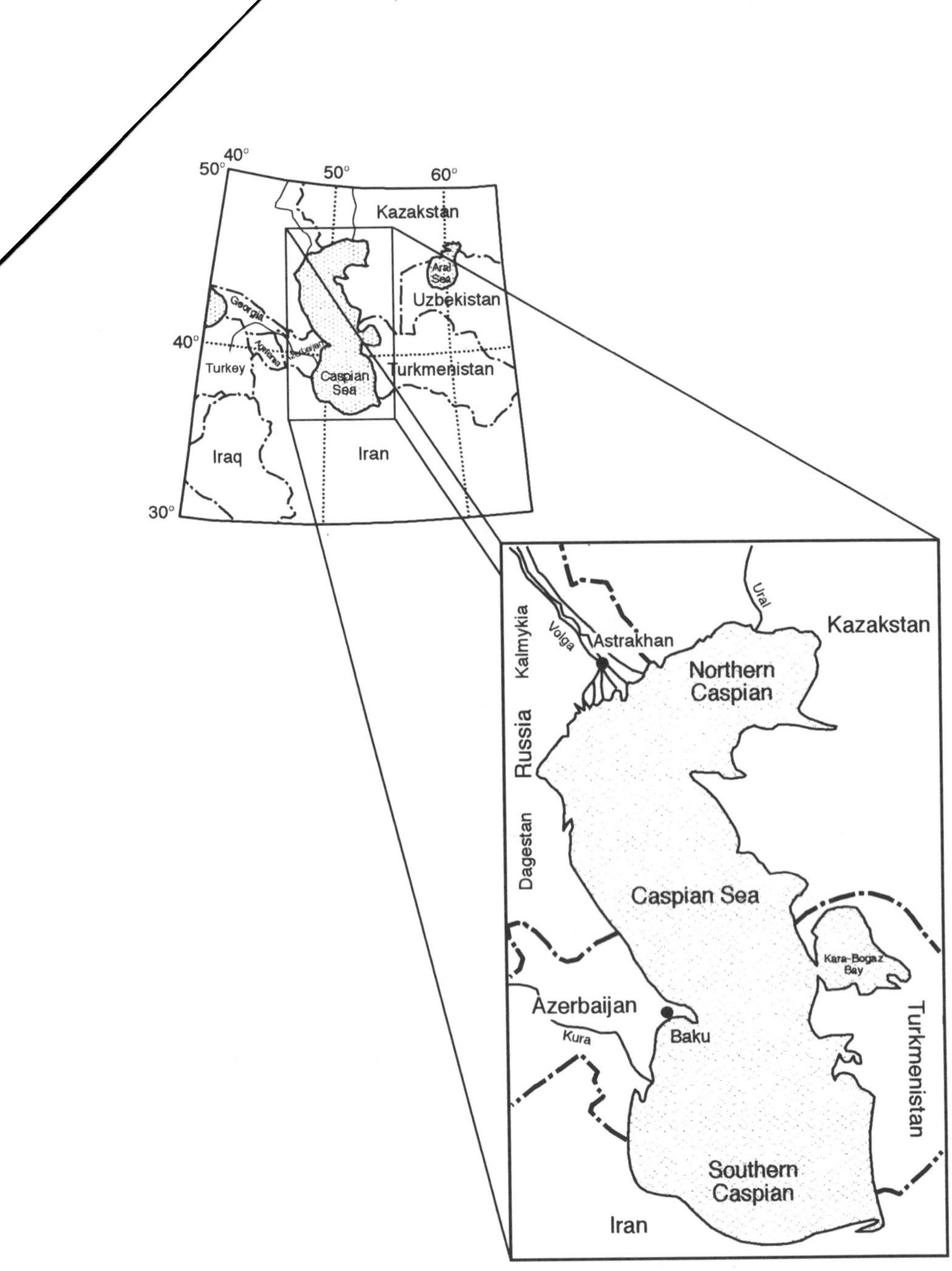

Fig. 1. The Caspian Sea

the catchment area of the Caspian Sea, supplies nearly 80% of the total volume of the sea (Fig. 2). All components of the Caspian ecosystem, directly or indirectly to a greater or lesser extent are influenced by river flow.

The Nature of Caspian Sea Level

Fluctuations in sea level of various duration can be found in the data of geomorphological and historical studies of the record of the Caspian Sea. Within the last 10,000 years, the amplitude of fluctuations of Caspian Sea level has reached 15 m (varying from −20 to −35 m). During the period of instrumental observations (from 1830), this value was 4 m, varying from −25.3 m during the 1880s to −29 m in 1977 (the lowest level occurring not only within the last 50 years but since 1830). Positive annual increments in the level during this period exceeded 30 cm on three occasions (in 1867, 38 cm; in 1979, 32 cm; in 1991, 39 cm). The mean annual increment in the level in the 1978-1991 period was 14.3 cm.

There are three widely known main hypotheses on the causes of annual fluctuations of Caspian Sea level:

1) *probabilistic*; considers fluctuations of sea level as a result of random variations of the various water balance components;

2) *tectonic*; associates fluctuations of the level with tectonic activity in the Caspian Sea region;

3) *climatic*; assumes that sea level fluctuations are the reflection of large-scale hydrometeorological processes, taking place not only in the sea's basin, but far beyond its borders [13].

The majority of researchers at present supports the climatic hypothesis.

Generalization of the results of comparing the changes of the Caspian Sea level and characteristics of Northern Hemisphere climate led to two conclusions [19]:

1. *The rise in level correlates well with the period of climate cooling*. Within limits, the variability of hydrometeorological processes increases both from year to year and from one run of years to another. Variability also increases with the weakening of the zonal circulation of the atmosphere over the North Atlantic and, most likely, over the whole Northern Hemisphere.

Since 1972, changes in the circulation regime have taken place, at first expressed in variations of cyclonic activity, which increased by 12% when compared to the previous decade. Especially sharp climatic changes in the region have been observed from 1976. There was an increase in the number of Atlantic cyclones (by 48%) and West European cyclones (by 31%) with a simultaneous increase in their water supply (by 35 and 18%, respectively). Against the background of a global increase in air temperature,

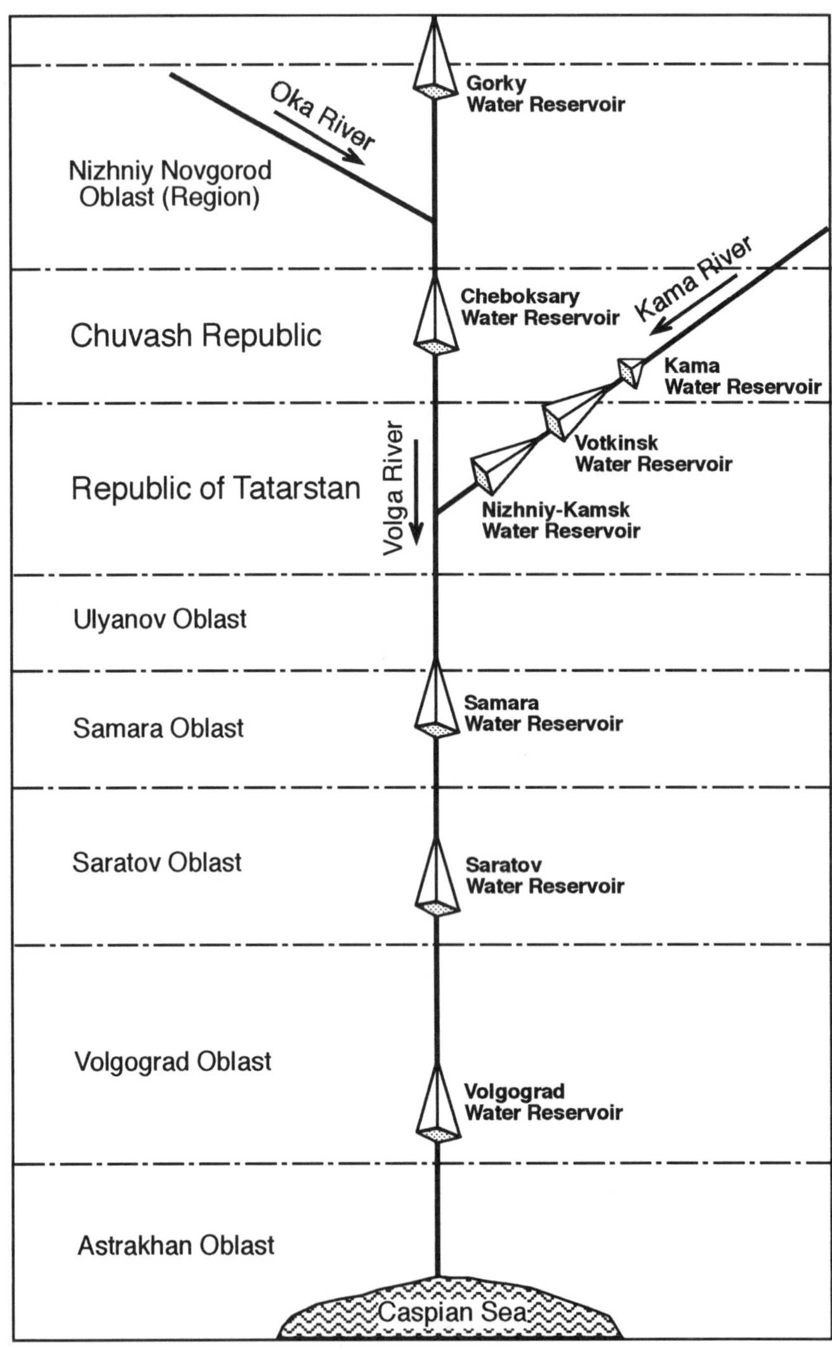

Fig. 2. Volga River Basin

this resulted in an increase in cloudiness, precipitation (and, as a consequence, river flow), and a reduction in evaporation (the reduction of the sea water temperature also had its effect on the decline of evaporation).

2. *A decline in Caspian Sea level accompanies climate warming*. The increase in air temperature, which will be noticeable in high latitudes, results in a weakening of the meridional temperature gradient. In this situation, the intensity of the zonal circulation of the atmosphere increases. The Icelandic Low and the Azores High pressure systems are intensified and shift to the north. As a result, the basic trajectories of cyclones are shifted northward, and a high pressure field with small amounts of precipitation is formed over the Volga River basin.

A comparison of weather forecasting data with the level of the Caspian Sea points to the reliable correlation between these processes. During the years of increased sea level, the number of rainy days significantly exceeded the number of dry days (especially during cold periods). Also, the reduction of "rainy weather" coincides with periods of decline in sea level.

The following periods can be identified in the annual fluctuations of the Caspian Sea (depending mainly on climatic factors) [15]:

1) 1900-1929 was a period of relative equilibrium of water balance components, the sea level fluctuated insignificantly around the actual sea level of −26.2 m;

2) 1930-1941 was a period of extreme deficit vis à vis the water balance (62 km^3), accompanied by a disastrous drop in sea level by 1.8 m (16 cm/year on average); this was the result of the reduction of the river flow (mainly of the Volga River) and intense evaporation from the sea surface;

3) 1942-1977 was a period of a moderate deficit in the water balance (13.7 km^3/year on average) and slow decline of the sea level (by 3.7 cm/year);

4) 1978 to the present (1996) is a period of a positive water balance and an abrupt rise of the sea level from its lowest mark of −29 m in 1977 to the levels that existed at the end of the 1930s and the beginning of the 1940s (Table 1); in 1990 the mean annual level increased to −27.5 m.

Studies by Golitsyn et al. [7, 8] indicate that the changes of the Caspian Sea level within recent years are almost 90% associated with corresponding changes in the water balance components of the sea as opposed to tectonic activity. The volume of surface inflow to the sea increased sharply after 1978. During certain years (1979, 1985 and 1990), more than 350 km^3 of river water entered the sea. From 1978 until 1990, the Volga River flow exceeded 260 km^3/year.

TABLE 1. Data on the drainless depressions near the Caspian Sea coast [11]

Depression	Level of bottom (m)	Area km^2 (x 1000)	Volume (km^3)	Distance from Caspian Sea (km)	Height of watershed (m)	Volume of water from Caspian Sea (km^3/year)
Kara-Bogaz-Gol Bay (Turkmenistan)	−32	18	120	11	0	20
Sory Mertvyj Kultuk, Kajdak (Kazakstan)	−27	10	10	0	1	11
Karagie (Kazakstan)	−132	1	92	21	7	2
Kaundy (Kazakstan)	−55	0.7	7	22	100	1
Zhazgurly, Basgurly (Kazakstan)	−26	0.3	2	30	130	1

Atmospheric precipitation has also contributed to sea level rise. Its amount between 1978 and 1990 significantly exceeded the annual rate, reaching 230 mm/year. Thus, more than 52 km^3 of water entered the sea compared with the amount used for flow and evaporation into the Kara-Bogaz-Gol Bay. This resulted in positive increments in the level: an increase of 14 cm of water per year and up to 30 cm in certain years. The evaporation from the sea's surface, according to certain assessments, could have contributed to the rise in sea level by more than 30%.

The damming of the Kara-Bogaz-Gol Bay in 1980 in order to maintain the Caspian Sea level also contributed (3 cm/year on average) to sea level rise, which from 1980 to 1990 was 35-40 cm, i.e., nearly 25% of the total value of the increase in that decade [15]. As one of the causes of the "unexpected" rise in Caspian Sea level, Tursunov [24] suggested the idea that the salt and dust clouds of Aral Sea dust and salts moving toward the Caspian Sea, blocked the sun's rays over the sea and its coastal regions, thereby increasing precipitation and reducing evaporation.

At present, there are no arguments that can challenge the view that the main contribution to seasonal and annual level fluctuations of the Caspian is accounted for by surface inflow and evaporation levels. Within recent decades its fluctuations have been subjected to anthropogenic impacts as well.

The reduction of inflow to the Caspian, as a result of human economic activity was, for example, in 1990, 41 km^3. Calculations show that, if there had been no increase in river water use for economic growth of the countries of the basin, the level of the sea would be 1.2 to 1.3 m higher than today, and its decline would have stopped by the end of the 1950s.

Current tectonic movements of the earth's crust on the Caspian Sea coast do not exceed several mm per year, and their impact on the variability of sea level can be considered insignificant [15]. Recently, a tectonic hypothesis was promoted, explaining the inflow and outflow of waters of the Caspian Sea, as a result of compressions and strains of the earth's crust [22, 10]. In particular, it has been suggested that almost all errors of the water balance equation are connected with earthquakes of a certain magnitude. During the periods of lengthy contractions of the earth's crust the whole reserve of water, contained in its pores, is released to the sea and its basin is overfilled. With extending stresses or even with the relaxation of a contraction, water leaves the sea, as it is absorbed into pores of the earth's crust.

Iranian specialists registered high tectonic activity in the regions of Mazendaran and Albortz. They evaluated the rate of vertical tectonic movements to be 3-5 cm/year (in the region of Apsheron Peninsular −3 cm/year). Also noted was the origination of small islands on the transect between Baku and Krasnovodsk (now called Turkmenbashi) [1].

As Babaev and Kubasov noted [2], the rise of the earth's surface near the location of earthquakes is a general feature of the processes in the development of strong earthquakes. Such a rise occurs along the whole extent of the Kopetdag-Caucasus seismic zone. It was noticed before the Ashkhabad (now spelled Ashgabat) earthquakes in 1948 and 1978, and it was monitored before the Spitak earthquake in 1988 and the earthquake in South Ossetia in 1990. Data about the repeated leveling at the Nebitdag geodynamic polygon suggest the possible rise in the the bottom of the southern Caspian Sea; following this, there is a systematic increase in the rate of modern vertical movements of the earth's surface (uplift) from Kumdag toward the coast of the Caspian Sea.

The tectonic connection between the Aral and southern Caspian Seas has been suggested by some scientists. Kochemasov [12] believes that alternating draining-filling of the linked system of the Aral-Caspian seas takes place at various periods. Golubov [9] presumes that underground nuclear explosions on Mangashlak, having disturbed the integrity of the regional impervious bed of Paleogene clays, could not only increase groundwater flows to the Caspian, but could also provide drainage of surface waters of the Aral Sea into the Caspian Sea basin. It is important to mention the Azgir bridgehead which is located in the Atraus region, where more than 9 nuclear blasts were detonated in salt domes [25].

The future preservation of current components of the water balance of the Caspian may cause its level to rise to −24 m. Based on an analysis of current water-exchange processes using isotopic data, Ferronskii et al. [5] concluded that the process of stabilization of the Caspian Sea level will start in 1996-1997 and will be completed in 12-15 years. Following stabilization, they suggest that a period of the sea's recession will begin.

Most researchers agree that the rise of the Caspian will continue at least until 2010. Up to then the synoptic circulation pattern, having begun in the 1970s, when the trend changed from decline to increase, should continue. In this scenario its level will reach −25 m within the next decade.

Rychagov [21], when forecasting the level regime of the Caspian Sea on the basis of paleogeographic reconstruction, comes to a conclusion that, under current physical and geographic conditions typical for the sub-Atlantic epoch of the Holocene (i.e., within the last 2,000-2,500 years), the level of the sea has not exceeded −35 m. This suggests that the rise of the Caspian will not exceed this value, and that taking into account economic activity and the restoration of flow into the Kara-Bogaz-Gol Bay, the level will come to rest at −26 m. However, according to the data of Arlamadkhan [1], if the rate of increase of 15.6 cm/year (1994) is maintained, the level may reach the −22 m mark within the next 25 years.

According to the forecasts of Osika [16], the rise of the sea to −26 to −25 m by 2005-2010 is possible, and to the −24 m level by 2020. This will be followed by a period of quasi-stabilization for 30 to 40 years, and again its level will decline to −27 to −28 m.

There are forecasts, according to which the level of the Caspian Sea may continue to rise until 2035 reaching the −21 m level and above. In this connection the question arises as to how precisely the amplitude and trend of changes of the sea level can be predicted. Analysis of available data makes it possible to conclude that at present there are no reliable forecasts. Therefore, at present only the hypothesis of quasi-cyclic fluctuations of the Caspian Sea level can be justified.

Thus, the level of the sea in the future will be subjected to annual fluctuations and decadal scale between the −20 m and −26 m marks (the levels prior to 1929) and −29 to −33 m which is close to the level (and therefore) coastline of 1977. Stabilization of the rise of the Caspian Sea level is also possible at −26 m to −25 m; it was at these levels for a significant period of time in the 19th and the beginning of the 20th centuries.

When discussing Caspian Sea level rise, one cannot help but note the impact of wind-induced wave surges on the coast. Especially large surges are typical for the northwest Russian coast. They are sparked by prevailing east and southeast storm winds (mainly during cold periods of a year). As a rule, they last for several days, reaching a height of 2.5-3.0 m against the background of annual inundation of 0.1-0.3 m. A large number of major surges (with heights of 1.5-3.0 m) was registered in the Caspian Sea during the last 120 years. The largest one took place on 10-13 November 1952 when a strip of land 25-50 km in width was inundated. At certain locations the water was 3 to 4.5 m deep. Similar wind and wave surges took place in March 1995 along the Kalmykia and Dagestan coasts, resulting in the death of people and significant material losses comparable to the total losses for all years of rising sea level [18].

Social, Ecological and Economic Consequences of the Caspian Sea Level Rise

Apart from the inundation of vast areas, the rise of the Caspian Sea level has been responsible for the following: groundwater level increase, intensification of adverse wind and wave impacts along the coast, erosion of exposed and submerged coastal slopes, increased (deeper) penetration on land of water surges, and the intensification of seismic activity. Of no lesser hazard are the synergistic effects connected with the rise of sea level which acts as a barrier to river flow into the sea and to the accumulation of bed-load material at the mouths of large rivers. This has already created conditions for disastrous breaches of the Terek and Sulak channels, and has significantly aggravated navigation conditions in the Volga delta.

Also synergistic in nature, and intensifying each year, are the processes of salinization of the inundated lands, salinization and pollution of groundwater, and the entrainment of oil products and various toxic substances into the sea, including wastewater from destroyed main drains [18].

The Russian coast of the Caspian is occupied by the territories of the republics of Dagestan and Kalmykia and the Astrakhan Oblast. Their coastlines are 490, 100 and 95 km, respectively, for a total of 685 km.

In the Astrakhan Oblast, 10% of the territory went out of agricultural use by 1995. More than 400,000 ha of valuable coastal areas have been inundated, and 12,000 ha of especially valued irrigated cropland also went under water. The Caspian Sea reached the onshore oil fields of Martyshi where the water is already a half-meter deep. Forty-five settlements have been inundated (27 of them to such an extent that they require relocation), along with dozens of kilometers of railway and automobile roads, power transmission lines, and considerable infrastructure. From the 1980s, there has been an intensification of the overflow from navigation and fish-pass canals, cutting through the shallow coastal waters of the Volga River mouth. In the lower part of the Ural River delta there is a strip of 10-12 km that is already continuously inundated [5]. The direct damage during only 18 years of rising sea level has cost more than $5.5 billion, and it is increasing exponentially [17].

The coastline of the Republic of Dagestan, which had become dry land due to the long-term decline of the Caspian Sea level, became the arena for intense agricultural development and spontaneous migration of the mountain population (at present, 40% of population of the Republic resides along the coastal region of Dagestan). Today there are 40 production facilities, partially inundated, in the cities of Makhachkala, Kaspiisk, Derbent and in the Sulak settlement. More than 800 families have been resettled. Nearly 150,000 hectares of land have been inundated, where hundreds of facilities for distant livestock breeding are located. The normal functioning of drainage systems has been disturbed, leading to groundwater rise, which adversely affects the yields of agricultural crops [6].

In the Republic of Kalmykia, the sea approached near the city of Legan and the settlements of Severnyi, Krasinskii, Dzhalykovo, and Burannoe. 7,000 ha of agricultural lands have been adversely affected by inundation and groundwater rise. Many of these lands were most valuable for forage production [4].

Much of the coast of the Republic of Turkmenistan (650 km in length) is made up of sandy beaches and dunes. A significant part of the structures of the coastal settlements is built on such formations. In connection with the encroachment of the Caspian Sea, sand formations are easily susceptible to destruction. Thus, the recreation area of Kheles is already under water. The Cheleken peninsula has become an island separated from the mainland (see the Babaev chapter, this volume). A threat hangs over the port and moorage of Krasnovodsk Bay. With a further rise of the sea level, western Turkmenistan will be flooded: the lower settlement of Cheleken, the ports of Alatepe, Aladja, the recreation area of Avaza, etc. The city of Turkmenbashi (formerly Krasnovodsk) is most significantly endangered, where communications of oil products, supply structures of the Krasnovodsk Oil Refinery, a pumping station, and part of the Ufta settlement are in various stages of flooding.

In Kazakstan, where the coastline is 1,600 km long, the advance of the sea (with a level rise to −27 m), has already caused the loss of 357,000 ha of agricultural lands; 200,000 head of livestock have died due to the lack of forage.

In Azerbaijan, where the coastline is 850 km long, water has approached Sumgait; Kuril'skaya Bay bar and the Sara peninsula have been transformed into islands. In Lenkoran the Caspian Sea has already inundated more than 800 private houses, 1,000 ha of fertile land, and the unique Kazylagachskii Reserve.

In the Islamic Republic of Iran the provinces Mazandoran and Gilyan experience the greatest impact of the rise of the Caspian Sea. The rise of the sea level by 2 m has caused damage to the coastal cities of Babolser, Tonkabon, Ramsar, Ashuradekh, Bender-Torkemen, Enzeli, Astare and Kolachai. Three thousand families were resettled from the coast [14].

According to the forecasts of continued rise of the Caspian Sea up to −25 m, its zone of impact would cover the lands of Dagestan, Kalmykia and Astrakhan Oblast to a total area of 1,654,700 ha, of which 1,001,700 ha will be inundated and 653,000 ha affected by an increase in the level of groundwater. In this case the inundation zone would include 53 cities and urban settlements with a population of 58,500 and 61 rural settlements with 416,000 people; 479,8000 ha of agricultural lands, of which 40,500 ha would have been irrigated; 384,500 km of highways, and the facilities for power generation, transport and communication, recreation and other branches of the national economy [23].

In Kazakstan the length of the coastline with the rise of level up to −26 and −25 m would increase to 2400 and 2700 km, respectively; the inundated strip of coastal land would occupy 1-3 km of the Mangyshlak coast and up to 25-30 km on the northwest coast. The total area of inundation with a sea level of −26 m would be 1.2 million ha (2.3% of the total territory of the Atrauskaya Oblast and 5.6% of Mangistauskaya Oblast; with the level of −25 m there would be an inundation of up to 2.2 million ha of the territory. Nearly 300,000 people and large industrial, municipal and agricultural facilities are in this zone. The adverse impacts of wind and wave surges along the coast should also be taken into consideration.

Impacts of the Sea Level Regime

With the rise of the Caspian Sea above a level of −26 to −25 m, there will arise the need for improved regime management of the sea. Provisional calculations show that excessive water mass entering the Caspian Sea and causing the rise of its level contributes about 40 km^3 to the sea. This volume of water can be withdrawn by increased water use or by water diversions into drainless depressions and other basins (Table 1). The natural outflow of the water into drainless depressions is a kind of self-regulating mechanism of the Caspian Sea, which "operates" under both low and high sea levels.

Under low sea level conditions, the main outflow of Caspian Sea water is directed to the Kara-Bogaz-Gol Bay. In 1977 when the sea level was at −29 m, nearly 5 km^3/year of water entered this bay. In 1980 flow to the bay was shut off by a fixed dam, but then the flow was restored partially (up to 2 km^3/year). In 1992 the government of Turkmenistan reestablished the natural flow into this bay, because of Caspian Sea level rise (within only 9 months, nearly 18 km^3 of water were supplied to the bay). If the sea level increased to −26 m, up to 20 km^3/year would be supplied to the Kara-Bogaz-Gol Bay. However, the assessments show that the bay's intake is limited to a volume of 120 km^3 [3]. In this case the ecological consequences of the increase of water surface area are not clear.

With a further rise of the Caspian Sea level under natural conditions, the second stage of self-regulation of the sea will begin to "operate" due to water outflow (up to 11 km^3/year) into the playas of Mertvyi Kultuk and Kaidak in Kazakstan. The cumulative impact of drainless depressions may result in the stabilization of sea level at −26 to −25 m, the same level which was observed at the turn of the 20th century.

Further studies are needed to analyze the water balance of the Caspian Sea under its current and future high levels and for critical analyses of the alternative optimization of water management activities.

References

1. Arlamadkhan, B., 1995: Problem of Caspian Sea Level Rise. *Proceedings of the International Scientific Conference: Caspian Region: Economics, Ecology, Mineral Resources*, held in Moscow, Russia, 20-23 June 1995, p. 8. Moscow, Russia: Ingeocenter. (In Russian)

2. Babaev, A.G., and I.M. Kobasov, 1993: Ecological problems of the Caspian Sea. *Problems of Desert Development*, **2**, 3-7.

3. Babaev, A.G., K.N. Amanniyazov, and V.P. Fedin, 1994: Transgression of the Caspian Sea and ways of its overcoming. *Problems of Desert Development*, **1**, 3-11.

4. Bogdanov, V.P., 1994: Caspian coast protection on Kalmykian territory. *Melioratsia I Vodnoe Khozyaystvo*, **1**, 7-8. (In Russian)

5. Ferronskii, V.I., V.S. Brezgunov, V.V. Romanov, L.S. Vlasova, V.A. Polyakov, A.F. Bobkov, and K. Frelikh, 1995: Study of the nature and mechanism of Caspian Sea level variations using isotope content of its water and bottom sediments. *Proceedings, International Scientific Conference on Caspian Region: Economics, Ecology, Mineral Resources*, held in Moscow, Russia, 20-23 June 1995. p. 4. Moscow, Russia: Ingeocenter. (In Russian)

6. Gladysh, L.N., and B.D. Tagirov, 1994: The state and problems of the protection of the Caspian coast in Dagestan. *Melioratsia I Vodnoe Khozyaystvo*, **1**, 5-6. (In Russian)

7. Golitsyn, G.S., 1989: Once more about the changes of the Caspian Sea level. *Vestnik AN SSSR*, **9**, 59-63.

8. Golitsyn, G.S., and G.A. McBean, 1992: Changes of the atmosphere and climate. *Proceedings of the Russian Academy of Sciences*, Geographical Series, **2**, 33-43.

9. Golubov, B.N., 1995: Abnormal level increase of the Caspian Sea as a result of long-term exploitation of the region. *Proceedings of the International Scientific Conference: Caspian Region: Economics, Ecology, Mineral Resources*, held in Moscow, Russia, 20-23 June 1995, p. 9-10. Moscow, Russia: Ingeocenter. (In Russian)

10. Kaftan, V.I., 1995: Analysis of the cosmogeophysical process periodicities and Caspian Sea level changes. *Proceedings of the International Scientific Conference: Caspian Region: Economics, Ecology, Mineral Resources*, held in Moscow, Russia, 20-23 June 1995, p. 14. Moscow, Russia: Ingeocenter. (In Russian)

11. Kaplin, P.A., and E.N. Ignatov, 1993: *Kaspiy*, Technico-economical report. Moscow, Russia: Moscow State University.

12. Kochemasov, G.G., 1995: Techtonic connection between the Aral Sea and southern part of the Caspian Sea within superstructure of the East European crater. *Proceedings of the International Scientific Conference: Caspian Region: Economics, Ecology, Mineral Resources*, held in Moscow, Russia, 20-23 June 1995, p. 33. Moscow, Russia: Ingeocenter. (In Russian)

13. Kuksa, V.I., 1994: The southern seas: Aral, Caspian, Azov and Black, under anthropogenic stress conditions. St. Petersburg: *Gidrometeoizdat*, 74-150. (In Russian)

14. Mansuri, A., 1994: The Caspian Sea coast protection: Actual international problem. *Melioratsia i Vodnoe Khozyaystvo*, **1**, p. 9. (In Russian)

15. Nikonova, R.E., 1991: Long-term changes of the Caspian Sea water balance components and its role in the level fluctuations. *Proceedings of the All-Union Conference on the Caspian Sea Problems,* Convened in Gur'ev, 3-5 June 1991, 26-30. Gur'ev: Ingeocenter. (In Russian)

16. Osika, D.G., 1995: On the level regime of the Caspian Sea and its forecast for the near future. *Proceedings of the International Scientific Conference: Caspian Region: Economics, Ecology, Mineral Resources*, held in Moscow, Russia, 20-23 June 1995, pp. 17-18. Moscow, Russia: Ingeocenter. (In Russian)

17. Ragozin, A.L., Yu.S. Grebnev, and V.A. Pyrchenko, 1995: History and urgent problems of engineering protection of the Caspian coast of Russia. *Proceedings of the International Scientific Conference: Caspian Region: Economics, Ecology, Mineral Resources*, held in Moscow, Russia, 20-23 June 1995, pp. 121-122. Moscow Russia: Ingeocenter. (In Russian)

18. Ragozin, A.L., 1995: Synergistic effects and consequences of the Caspian Sea level rise. *Proceedings of the International Scientific Conference: Caspian Region: Economics, Ecology, Mineral Resources,* held in Moscow, Russia, 20-23 June 1995, pp. 120-121. Moscow, Russia: Ingeocenter. (In Russian)

19. Rodionov, S.N., 1991: Main stages of the Caspian Sea level dynamics and their dependence on climate changes during the last millennium. *Proceedings of the State Oceanographic Institute (GOIN)*, **183**, 24-36. Moscow, Russia: GOIN.

20. Rosenberg, I., 1996: Catching sturgeon. *Itogi Magazine*, 4 June, 48-50. (In Russian)

21. Rychagov, G.I., 1995: Level regime of the Caspian Sea and possibility of its forecast on the basis of paleogeographical reconstructions. *Proceedings of the International Scientific Conference: Caspian Region: Economics, Ecology, Mineral Resources*, held in Moscow, Russia, 20-23 June 1995, p. 6. Moscow, Russia: Ingeocenter. (In Russian)

22. Shilo, N.A., and M.I. Krivoshey, 1989: Connection of the Caspian Sea level fluctuations with earth crust tensions. *Vestnik AN SSSR*, **6**, 83-90.

23. TED (Technical-Economic Report) "Kaspiy", 1993: *Melioratsia i Vodnoe Khozyaystvo*, **3**, 2-5. (In Russian)

24. Tursunov, A.A., 1995: On forecasting climatic changes in Central Asia. *Problems of Desert Development*, **5**, 3-21.

25. *Nezavisimaya Gazeta* (Russian periodical), 1993: Nuclear blasts detonated in salt domes. 26 May 1993. Moscow, Russia. (In Russian)

ENVIRONMENTAL PROBLEMS OF THE CASPIAN SEA REGION AND THE CONFLICT OF NATIONAL PRIORITIES

TATYANA A. SAIKO
Department of Geographical Sciences
Faculty of Science
University of Plymouth, United Kingdom

Introduction

The Caspian Sea is currently one of the few locations in the world which attracts the attention of scientists, politicians, and businessmen. It is in many ways a unique region and can be considered as a model area for the study of the effects on society of a possible future sea level rise in connection with global climatic change. This experience is particularly valuable for countries, especially "island nations" such as the United Kingdom.

The Caspian Sea basin is one of the world's largest closed drainage basins and the sea is actually the greatest brackish lake on the planet. It is endowed with immense natural wealth in the form of rich hydrocarbon resources and 90% of the world's most precious sturgeon catch. The hydrocarbon reserves discovered relatively recently have been at the core of heated political debates and the largest business deals in the last few years. These resources are particularly valuable, as they are located outside the OPEC cartel which was responsible for the two oil shocks of the 1970s. The circum-Caspian region may easily become "the Kuwait of the Third Millennium" and may contribute to the economic development of the five riparian countries bordering the sea: Azerbaijan, Iran, Kazakstan, the Russian Federation and Turkmenistan. The interests of many more countries are focused on this region thus making it of global rather than only of regional importance.

Unless a properly coordinated monitoring system is introduced and agreed-opon environmental management and conservation strategies brought into practice, the expected multi-fold growth of oil production in the region may result in the complete disaster of another valuable resource — the stocks of sturgeon species. Environmental problems, already serious in the region, are being compounded by the continuing rise in the sea level, which started in 1978. There are certain political, economic and social factors which can contribute to the deterioration of the current

M.H. Glantz and I.S. Zonn (eds.),
Scientific, Environmental, and Political Issues in the Circum-Caspian Region, 41-52.
© 1997 *Kluwer Academic Publishers. Printed in the Netherlands.*

situation and limit the potential for finding a common program of action suitable for all countries. It appears that the present economic difficulties experienced by all littoral states are responsible for the current lack of effectiveness in environmental protection in the circum-Caspian region.

Current Environmental Problems of the Caspian Region and the Impact of the Rise in Sea Level

Environmental problems in the circum-Caspian region are numerous and diverse by nature, because of the variety of landscapes involved and the multitude of economic activities carried out in different parts of the drainage basin. The set of problems involves issues related to the sea proper, to the surrounding region and also to the more distant areas sharing the Volga River drainage basin. They include, but are not limited to, the following: water pollution by oil and other industrial, agricultural and domestic wastes; air pollution; eutrophication of the Volga delta; land degradation of the coastal territories; desertification in the Kalmykian steppes; depletion of commercial fish stocks throughout the sea. These issues are further affected by the rise in sea level.

There is a complex of interrelated pollution problems in the Caspian Sea basin. The main sources of pollution in the circum-Caspian region are: discharge of inadequately treated wastes of different origin into the sea and its tributaries; technical accidents at coastal enterprises and within the territorial waters of the coastal states; exploration, production and transportation of offshore and near-coastal oil and its products; surface runoff by storm waves particularly linked with increased sea level processes, e.g., from inundated oil and gas fields, etc. Surface river runoff is responsible for 90% of the total pollution load [5].

Serious problems associated with degradation of ecosystems in the Northern Caspian are primarily the result of pollution from the Volga River, which accounts for 80% of the inflow to the sea. The Volga is Europe's longest river with its drainage basin totalling 1.36 million km^2 (or 62% of European Russia). 150,000 water bodies including 2,600 rivers are part of this extensive river system. Annually, it brings 240 km^3 of water into the Caspian Sea. In 1992 the Volga region as well as the coastal area of the Northern Caspian and Kalmykia were designated as "ecological disaster zones" [16] due to the massive degradation of ecosystems within the basin. This enabled some experts to call it "a united front of ecological degradation between Scandinavia and the Black Sea" [24].

The Volga basin accounts for approximately 20% of all water pollution in Russia. Annually about 12 cubic km of polluted industrial, agricultural and domestic wastes are discharged into the Volga along its course [16]. For dilution, they require about 200-240 cubic km or almost all the river flow. In 1992 they contained 387,000 metric tons of organic materials, 396,000 metric tons of suspended matter and considerable quantities of phenols, pesticides and heavy metals [16]. According to estimates, 13,000 metric tons of oil and petroleum products pollute the river annually with a large share reaching the delta and the northern portion of the Caspian Sea. Soil erosion is responsible for much of the pollution load, as it affects approximately 7% of the whole basin area. In the wooded steppe zone 60-70% of humus is lost from the soils, while further north in the Central Region the humus layer is completely lost over an area of 1 million hectares.

Over the past decade there have been increased inputs of nitrogen, phosphorus and organic compounds, 30 to 40% of which come from anthropogenic sources. In some years they increase up to 60%. Processes of eutrophication have accelerated since the late 1970s in the river's reservoirs, the Volga delta and northern Caspian Sea [8]. It has been shown that the discharge of organic matter from the Volga to the Caspian Sea has recently increased (since 1978) because of enhanced water exchange as a result of the rise in the water level. As a consequence, oxygen deficits are a common feature in the western portion of the northern Caspian where, by 1990, 50% of the aquatic system was affected by oxygen deficit [15]. It is confined to water less than 1 m deep, but periodically extends into water 3-5 m deep. The overall quality of water in the Volga delta has deteriorated over the past decades and cyanobacterial blooms (eutrophication with the wide spreading of bacteria-producing cyanide substances) are now common in these areas.

Construction of hydroelectric plants and dams on the Volga River as well as pollution has been responsible for the decline in sturgeon stocks over recent decades. About 65% of the sturgeon catch in the Caspian is associated with the Volga, including 97% of the *Acipenser sturio* (Russian sturgeon), 66% of the *Huso huso* (Beluga sturgeon), and 25% of the *Acipenser stellatus* (Stellate sturgeon) [8]. After construction of the Volga hydroelectric power plant, only 430 ha out of the former 3,600 ha of the spawning grounds remained accessible to fish. Further, changes in the hydrological regime have caused a delay in the spawning of the Beluga, and its reproduction suffered. Despite the release of fry from hatcheries, there has been an overall decline in stocks over the last decade. As a consequence of water pollution, there have been reproduction problems in mature fish. Adverse toxicological conditions in the river have also resulted in a disease that manifests itself in delamination of muscular tissue and a weakening of egg shells, which affects the quality of caviar.

Oil pollution is already a serious problem throughout the Caspian Sea. An appraisal of the overall status of water quality in the Caspian within the former Soviet Union was made by Mnatsakanian [17]. According to his estimates in 1992, throughout the aquatic system the average level of pollution with oil products and phenols exceeded the maximum permissible concentrations (MPC) by 2 to 43 times. The worst polluted areas were those near the Ural River delta in Kazakstan and along the Azerbaijan coast. The average pollution levels around the Apsheron peninsula reached 10 MPC for oil and 18 MPC for phenol products. Near the coast of

Kazakstan the average pollution levels in 1993 were 4 to 6 MPC [12]. Data from the Kazak sources [25] also reveal that the situation is as serious within its coastal territory. The authors note that the main form of pollution is associated with the exploitation of oil and gas deposits of which there are more than one hundred including the giant Tenghiz oil field, which has recoverable reserves estimated at 1 billion metric tons of oil [19]. The ecological situation is especially hazardous in the region of the Karachaganac oil and gas condensate complex [25]. Drilling works in the shelf zone present a particular hazard for marine life. Seismic exploration for oil and gas reserves already resulted in massive deaths of sturgeon in 1987[5].

The threat of oil pollution relates not only to the areas of offshore oil exploration and extraction or shipping across the Caspian Sea, but also occurs as a result of transportation across the coastal zones. In the Russian Federation alone, there were over 1,800 technical accidents in 1994, resulting in the discharge of different pollutants, with a total damage to the environment estimated at US$4.3 million [10]. Oil spillage, owing to technical accidents, are particularly dangerous for aquatic ecosystems, because in this case concentrations of pollutants may exceed the maximum permissible levels by hundreds and thousands of times and accumulate in the food chain affecting the sea's bioresources and fish stocks.

It is expected that with the rising sea level in the Northern Caspian, part of the oil and gas fields within the current coastal zone will be flooded, thus increasing the total amount of petroleum products reaching the open sea. It is estimated that by the year 2005, within the Russian territory alone, 8 towns and 105 villages with a total population of 197,000 people will be affected by inundation and flooding. More than 631,000 ha of agricultural lands may suffer, as well as the locations of 15 oil wells. Within the Russian territory, the most serious problems are expected to occur along the Dagestan coast where the main portion of deposits are located. In Kalmykia it is predicted that an increase to the minus 25 m level will completely cover the Kaspiyskoye oil and gas field [5]. In Kazakstan several major oil fields will be flooded as well as 594,000 hectares of agricultural lands and 30 settlements. In Turkmenistan 15 settlements, large animal farms, ports and most of its coastal oil and gas fields will be affected [14].

It is also expected that the Caspian Sea level rise might contribute to contamination of the shallow waters of the Volga delta with hydrogen sulfide, thus diminishing its productivity by 4 to 5 times and resulting in a decline in sturgeon catches by up to 1.5 times [23, 5]. According to estimates, stocks of sturgeon — the world's main source of caviar — have already reached precariously low levels during the last five years. Stocks have dropped from 200 million five years ago to no more than 60 million today [21]. Production of caviar in Russia has also fallen to only 70-90 metric tons in 1995 [3]. The structure of other fish stocks has also changed dramatically with an increase in abundance of low-value fish species [15].

Air pollution contributes to the overall pollution load. The majority of enterprises are not equipped with adequate treatment facilities. Normally 61-85% of solid particles are intercepted in Kalmykia and Dagestan as compared to the Russian average of 94%. Gaseous compounds are hardly intercepted at all and in 1992 all discharged sulfur dioxide and other gaseous pollutants in these regions reached the atmosphere [5].

Processes of land degradation and desertification which affect the coastal areas of the circum-Caspian region present a serious threat to the environments of Kalmykia and Turkmenistan. In Kalmykia overgrazing of steppes has resulted in the creation of the first anthropogenic desert in Europe [22, 27]. There has been some progress in the ecological situation. According to the Ministry of Ecology and Natural Resources, the total area of active desertification has decreased from 3,760 km^2 in 1985 to 2,780 km^2 in 1994 because of a number of stabilization measures taken within the republic. In 1994 alone US$3.3 million or 97% of all environmental expenditures had to be spent on land conservation and management [16]. According to estimates [27], almost 80% of the republic is still affected by the processes of desertification.

It is clear that disastrous consequences of the rise in sea level have resulted primarily from unsound economic development within the high-risk coastal zone during the sea's low-level period of 1930-1977. New settlements and industrial enterprises were placed there without due regard of the hazards involved. There have already been serious implications of the rise in sea level. Over 100,000 ha of land have been withdrawn from economic use because of the effect of inundation and flooding of the coastal zone within the Russian Federation. In the spring of 1995 it was reported that in Kalmykia two people had been killed, 900 evacuated and 441 homes were flooded, as a result of unusually severe storm winds hurling seawater at the coast. In Dagestan 81,500 sheep and cattle had died, 127 homes had been destroyed and 500 homes flooded [9].

Outmigration of people from these predominantly agrarian and already overpopulated regions has started. For instance, in Dagestan the total population of local national origin in 1979-1989 increased by 14.5% while outside the republic in Russia it had increased by 31% [5]. It should be emphasized that in cases of resettlement in response to the impacts of rise in sea level the affected population is not properly compensated for losses in housing and property. This can be further complicated by the land privatization campaign, which puts additional pressure on these multinational regions that are potentially quite vulnerable to ethnic and social tensions.

Implications of Current Economic Problems for Environmental Management and Control

Political aspects involve a number of issues which have become topical in the new geopolitical situation after the dissolution of the former Soviet Union. Discussion of various disagreements and sometimes even ethnic conflicts arising between the coastal states is not the focus of this chapter. However, it should be mentioned that to use and conserve Caspian resources effectively, it is essential to reach a consensus on the basic issue of determining the legal status of the Caspian Sea. The five riparian states have not yet agreed whether it should be regarded as an inland sea or a lake. Before the dissolution of the former Soviet Union, the legal status of the Caspian Sea was that of a lake and consequently it implied the common use of its resources by the USSR and Iran. After each former republic had become independent, its individual attitudes to this issue changed. Russia and Iran still regard it as a lake while Azerbaijan, Kazakstan and Turkmenistan consider it a sea, which assumes its sectoral division and independent use of natural resources.

These attitudes are clearly linked to such economic issues as the availability of offshore oil resources recently discovered near the coasts of the latter three countries. Immensely rich oil reserves of the Caspian are at the core of many other disputed political and economic issues. According to different estimates, from 40 to 57 billion barrels of oil reserves have been discovered under the seabed and in the surrounding region [1, 2]. Potential reserves are estimated at 68-250 billion barrels [26]. Three major deals related to offshore oil production at a total cost of US$13.3 billion, which for the above-mentioned reasons are considered illegal by Russia, have already been signed by Azerbaijan with international oil consortiums involving 16 companies, including Russia's "Lukoil."

Recently, Kazakstan signed an agreement with Russia for a Caspian Pipeline Consortium. A US$1.5 billion pipeline will export oil and gas mainly from the giant Tenghiz oil and gas field through Russia to the Black Sea, and then on to Western Europe and the rest of the world [2]. Estimated recoverable reserves of this oil field, developed mainly by the Chevron company, amount to 1 billion metric tons, while Kazakstan's offshore reserves are even 3.5 times greater [4]. Thus, it appears that the issue of the legal status of the Caspian Sea will possibly be solved only when each of the five states agrees on its share in the oil wealth of the region.

Despite many differences in the level and specific nature of economic development in each of the five countries, there are some similar features. All countries, including Iran, are undertaking economic reforms and they are experiencing serious economic difficulties. For the former Soviet republics these difficulties include high inflation rates and a continuous decline in overall economic production. The European Bank for Reconstruction and Development reckons that in real terms, Azerbaijan's GDP in 1995 was a mere 34% of its 1989 size. The figures for Kazakstan and Turkmenistan were 45% and 63%, respectively [2].

The Russian Federation had certain signs of economic stabilization during the last two years. However, Fig. 1 shows that in 1994 even Russia, with the greatest GDP of all the new East European democracies, had a per capita value of only half that in Slovenia and less than in Poland, a country which has only one-third of Russia's GDP. It also demonstrates that during 1990-94 all four former Soviet republics had a negative annual average change in the value of GDP ranging from minus 4 in Turkmenistan to minus 15 in Azerbaijan and an annual average increase of consumer prices exceeding 300% in Russia and Azerbaijan and 500% in Kazakstan and Turkmenistan [6]. There have been some positive changes in the economy of Kazakstan. In 1995 inflation in this country was 60.3% compared to 260% in 1994 [13].

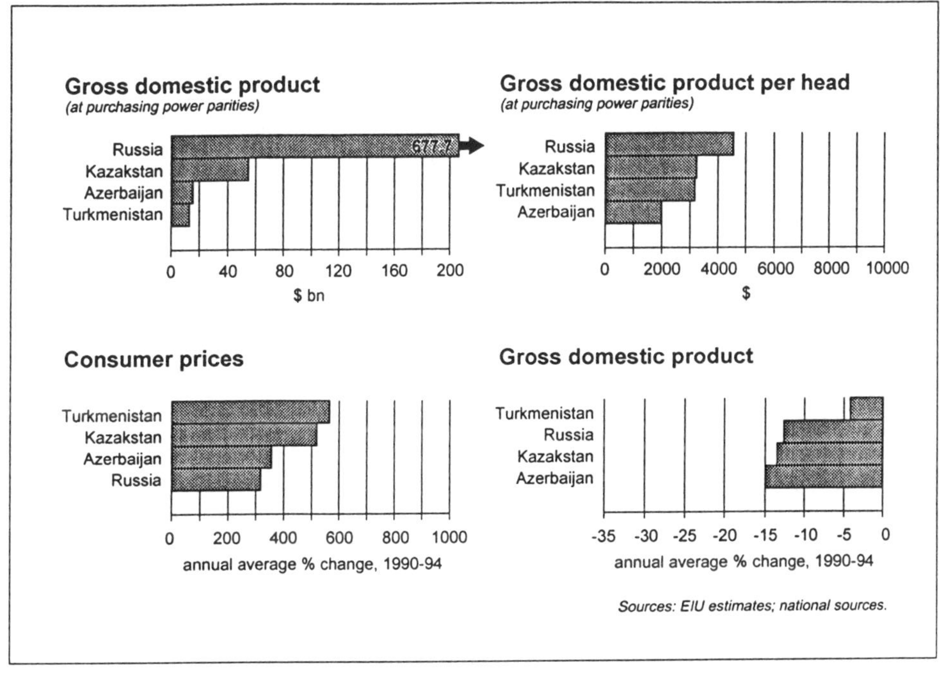

Fig. 1. Selected economic indices of former Soviet republics

Iran's per capita GDP in 1992 was only US$2,230; its economy never fully recovered from the effects of its Gulf War with Iraq. Although this country's GDP has been increasing at an estimated annual average rate of 2.3%, it has serious economic and social problems, such as high inflation exceeding 20% per year in 1992-1994 and high unemployment estimated at 10% in September 1993 [7].

All littoral countries except Russia are additionally affected by the problems associated with high rates of population growth. People living around the Caspian Sea tend to be much less concerned with environmental problems than with the everyday struggle for survival. For instance, recently the Turkmenistan population of the coastal region had serious flour and food shortages. They are distanced from the problems of the decline in sturgeon stocks also, because the majority have probably never eaten either the precious fish or caviar. The cost of 100 g of caviar is about US$40 while, for instance, in Turkmenistan in 1994 an average monthly salary was estimated at US$30 [6].

The economies of all five countries rely rather heavily on exports of oil, gas, and oil products. According to different sources, the share in total exports amounts to 90% in Iran [7]; 52.4% in Azerbaijan [6]; 41.1% in Russia [11]; 43.5% in Turkmenistan and 29.7% in Kazakstan [6]. For most countries oil is the only available source of currency needed to revive the economy and protect sovereignty in the future. The current rates of oil production in Kazakstan, Turkmenistan and Azerbaijan are relatively low, compared to potential reserves. It is expected that after the year 2000 Kazakstan's annual crude oil exports will increase from the 1995 value of 10.6 million metric tons [19] to around 70 million metric tons [20]. The annual oil production in Azerbaijan from the new oil fields will reach 5 million metric tons by the end of 1996 and rise to 45 million metric tons after the year 2000 [18].

This multifold increase in oil production will inevitably result in the incomparably higher levels of oil pollution in the Caspian Sea and the neighboring coastal regions at the expense of its biological diversity and ecosystem health. It is doubtful whether the countries with collapsed economies would be able to afford expensive oil-collecting facilities to be used in cases of oil spills or other technical accidents. It appears that sturgeon may lose again in the competition with industrial development like it did when the large-scale construction of hydroelectric plants and dams started on the Lower Volga and when extensive seismic oil exploration has resulted in large-scale fish deaths as mentioned above.

Currently there are many contradictions among the data on the level of oil pollution in each of the five states. One reason for this is that, before the dissolution of the former Soviet Union, environmental monitoring was centered around the USSR's Academy of Sciences and, after 1992, each country had to proceed on its own. For Turkmenistan, for example, it meant a start from almost zero and, as the exchange of information has almost stopped, it is now almost impossible to get comprehensive and objective information on the environmental situation. It is particularly difficult because, as a result of present economic problems, no country can afford to sustain a proper scientific fleet to monitor the pollution levels.

In 1994 expenditures on environmental conservation in the Russian Federation were 25% less than in 1993. Investments allocated for environmental conservation and management of natural resources made up only 3.9% of the total funds for industrial construction. In 1994 approximately US$3.6 million was spent on measures for water conservation and management, while it was estimated that the essential works needed for urgent rehabilitation of the polluted water bodies required at least three times more money [16].

Because of the overall decline in industrial production in the last few years there has been a minor decrease in total pollutant loads. However, while the overall decline in 1994 economic production in the Russian Federation amounted to 21% (compared to 1993), the total volume of air pollution decreased by only 12.7% and water pollution by 9% [16]. This demonstrates clearly that countries affected by the economic crisis would save on environmental expenditures. Normally, wealthier countries tend to spend more on environmental protection also, because in the developed countries the value placed on human health is much higher than in the countries of the former totalitarian regimes. In the poorer countries there seems to be a vicious downward spiral because of the interaction between the adverse state of the national economy and the state of the environment.

While the amount of total revenue of the Russian Federation from exports of oil, gas and oil products reached US$20.6 billion in 1994 [11], according to the Ministry of Ecology and Natural Resources, only US$1 million was spent on the construction of facilities and equipment for collection of oil products from the surface of rivers, ports and seas [16]. At the same time the total revenue from exports of all canned fish products, caviar included, was only US$27.6 million. It is clear from a national perspective that sturgeon would be easily sacrificed for the sake of getting higher oil outputs, even for the primary caviar-producing country — Russia.

These facts also explain why environmental matters are given the lowest possible priority in national investment programs. There is a conflict between the needs of economic revival and development and the need to spend funds on environmental management and conservation. This conflict of national priorities prevents the governments of most countries from the effective implementation of existing programs and investment in the new area of environmental protection and control.

There has been a positive shift from centrally funded to regionally or locally supported environmental control. Certain economic incentives including fees for pollution have been introduced, although not very effectively implemented in the last few years. A recent example of possible ways to use the economic mechanisms for environmental control is an "ecological" tax which was imposed recently by the Krasnodar Krai Duma, the regional legislature, on export and transit-shipping operations in Novorossiisk, the Black Sea destination port for Caspian oil. However, it seems that the potential source of major funds for environmental improvement should be sought from the oil industry, which currently gets the greatest revenues from exports of this potentially dangerous pollutant.

Conclusion

It appears that at present the importance of taking into consideration environmental problems has not been properly recognized in ranking priorities and assigning appropriate financial allocations by the littoral countries. This apparent neglect has resulted mainly from a deep economic crisis experienced by most of them. The present scope and magnitude of environmental problems already threatens both the Caspian Sea and the surrounding coastal territories. These problems are compounded by the continuing rise in sea level. Thus, further in-depth studies are essential to monitor the environmental changes occurring within the sea's basin and to identify possible future levels of the Caspian Sea. There is a need to study the magnitude of Caspian Sea pollution and the contents of chemical elements of technogenic origin.

Controversial approaches of different countries to the issue of the legal status of the sea compound the difficulty of developing an integral approach to the utilization, management and conservation of the region's natural resources. Current problems associated with the uncontrolled continuing discharge of inadequately treated wastes into the Caspian, offshore oil exploration and production, and also the poaching of precious sturgeon fish in the open sea waters result from this issue. The intergovernmental agreement on the preservation and rational use of bioresources of the Caspian Sea has not yet been signed by all the Caspian states.

The Caspian Sea is a shared water body, so it is essential to achieve genuine cooperation between all the states in order to alleviate the current problems associated with the pollution which threatens the Caspian ecosystem health and with rise in sea level. It appears that the establishment of a regional research center for the integrated study of the circum-Caspian region may be a relevant and timely measure, particularly essential *before* the countries find an eventual consensus upon the majority of issues in contention.

KEY ISSUES:

- The importance of the circum-Caspian region is global rather than regional, because the interests of a large number of countries are focused on its immense oil reserves.

- The continuing rise in sea level compounds the current environmental situation.

- Environmental problems of the region are complex, are already quite acute and will inevitably be aggravated by the expected multifold increase in oil production.

- The main reason why the littoral countries fail to control the present environmental problems is their serious economic crises.

- There is a conflict of national priorities between the needs of economic revival and the development and the needs of environmental management and protection.

- In the poor countries of the region there seems to be a strong linkage between the adverse state of the economy and the state of the environment.

References

1. Anon., *The Economist Newspaper*, 9 September 1995.

2. Anon., *The Economist Newspaper*, 4 May 1996.

3. Business World, 9 June 1995, p. 10 - section in the *Moscow Times* daily newspaper.

4. Business Moscow News, 19 April 1995 - special business section in *Moscow News* weekly newspaper.

5. Centre for International Projects, 1995: *National Report of the Russian Federation on: Consequences of Climatic Change in the Caspian Sea Region*. Draft. Moscow, Russia: Centre for International Projects. (In Russian)

6. EIU (Economist Intelligence Unit), 1996: *Country Reports for Azerbaijan, Kazakhstan and Turkmenistan*. First Quarter. London, UK: The Economist Intelligence Unit Unlimited.

7. *Europa World Yearbook*, 1995: Volume 1, 36th ed. London, UK: Europa Publishing Ltd.

8. Finlayson, C.M., Y.S. Chuikov, R.C. Prentice, and Fischer, W. (Eds)., 1993: *Biogeography of the Lower Volga, Russia: An Overview*. Slimbridge, UK: IWRB Special Publication No. 28.

9. *Guardian*, 16 March 1995, p. 9 - daily newspaper published in the U.K.

10. Goskomstat, 1995: *Russia in Figures 1995*. Moscow, Russia: Goskomstat of Russia. (In Russian)

11. Goskomstat, 1995: *Russian Statistical Yearbook 1995*. Moscow, Russia: Goskomstat of Russia. (In Russian)

12. *Izvestiya*, 6 April 1993 (daily newspaper published in Russia).

13. *Izvestiya*, 23 January 1996, p. 6 (daily newspaper published in Russia).

14. Kaplin, P.A., 1994: *The state of the environment in the Caspian Sea region under the conditions of the sea level rise*. Unpublished manuscript (on file with author).

15. Kuksa, V.I., 1994: *Southern Seas (Aral, Caspian, Azov and Black) Under the Impact of Anthropogenic Stress*. St. Petersburg, Russia: Gidrometeoizdat. (In Russian)

16. Ministry for Ecology and Natural Resources, 1993, 1995: *The State Reports: Status of the Environment in the Russian Federation*. Moscow, Russia: Ministry for Ecology and Natural Resources. (In Russian)

17. Mnatsakanian, R.A., 1992: *Environmental Legacy of the Former Soviet Republics*. Edinburgh, UK: University of Edinburgh.

18. *Monitor*. Jamestown Foundation, **1**, *130*, 3 November 1995. (Internet)

19. *Monitor*. Jamestown Foundation, **1**, *128*, 1 November 1995. (Internet)

20. *Moscow Times*, 20 January 1995, p.6 - Moscow daily newspaper.

21. Open Media Research Center, 1995.

22. Saiko, T.A., and I.S. Zonn, 1995: Europe's first desert. *Geographical Magazine*, **67**, 4, 24-26. (Reprinted in this volume)

23. Shlikhunov, V.M., 1994: The coastal sea level rise: Problems and solutions. *Report of the World Coastal Conference*, 1-5 November 1993, at The Hague, The Netherlands.

24. Stewart, J.M. (Ed.), 1992: *The Soviet Environment: Problems, Policies and Politics*. Cambridge, UK: Cambridge University Press.

25. Sultangazin, U.M., and T. Tsukatani, 1995: Modelling of the Kazakhstan economy and environment. Discussion Paper No. 416, *International Forum on Aral, Caspian, and Dead Seas* held 27-29 March 1995. Tokyo, Japan: United Nations University.

26. Vassilev, R., 1996: Oil diplomacy in the near abroad. *Prism*, Vol. 2, No. 9, Part 2. The Jamestown Foundation.

27. Zonn, I.S., 1995: Desertification in Russia: Problems and solutions. *Environmental Monitoring and Assessment*, **37**, 347-363.

TOWARD REGIONAL COOPERATION IN THE CASPIAN: A LEGAL PERSPECTIVE

SERGEI V. VINOGRADOV
Centre for Petroleum & Mineral Law & Policy
University of Dundee
Dundee, Scotland

Introduction

One of the most "spectacular" contests over the riches of the Caspian Sea is currently being waged by four former Soviet Union republics (Azerbaijan, Kazakstan, Russia and Turkmenistan) and Iran. (For a more detailed summary of the background to this controversy, see [40, 39, 15, 21].) The demise of the Soviet Union and emergence of new independent States have drastically changed the entire geopolitical situation in the Caspian region. The extant legal regime of the Caspian Sea, based on the outdated but still valid agreements between the former Soviet Union and Iran, is no longer sufficient to deal with the host of complex political, economic and environmental problems affecting the region. The situation is aggravated by unilateral claims of the riparian States to the sea. Although the coastal States are currently discussing how the regime of the Caspian Sea might be resolved, the dispute is not yet settled and the potential for growing international tension is evident.

The issue of control over huge petroleum resources, proven and potential, of the Caspian seabed is at the heart of the current controversy. Unless and until this issue is resolved, no comprehensive and viable international legal regime with respect to the Caspian Sea can be established. However, existing uncertainty concerning the future legal status of the sea should not deter coastal States from cooperation in other important areas requiring their urgent attention: sea level rise, protection of the marine environment and conservation of biological resources. On the contrary, such cooperative efforts, without prejudice to these States' positions on the question of the legal status, could facilitate solutions of the more economically and politically sensitive issues.

Recent Regional Developments

Initial steps of the coastal States with respect to the Caspian Sea problems in a radically new political situation were encouraging. A meeting of experts from all littoral States convened in Tehran in October 1992 discussed Iran's proposal to create a regional orga-

M.H. Glantz and I.S. Zonn (eds.),
Scientific, Environmental, and Political Issues in the Circum-Caspian Region, 53-66.
© 1997 *Kluwer Academic Publishers. Printed in the Netherlands.*

nization with a broad scope of responsibilities including fisheries, exploration of oil and gas, transport, and prevention of marine pollution. The Conference launched the preparatory work on the Treaty on Regional Cooperation.

In October 1993, a meeting of the Heads of Government of Azerbaijan, Kazakstan, Russia and Turkmenistan confirmed that "the comprehensive solution of the problem of rational utilization of the Caspian Sea requires the participation of all Caspian States" [7]. Several areas of "joint activities" were identified, including:

- protection of natural reserves and natural resources of the Caspian Sea;
- conservation and optimal utilization of the biological resources;
- development of mineral resources with due account of economic interests of the Parties;
- determination of rational sea lanes with consideration of environmental concerns; and
- control of the level of the sea.

Meanwhile, the draft Treaty prepared by Iran in consultation with other coastal States was completed in 1993. The draft called for cooperation of the Caspian States in the utilization of the sea, which was viewed as a unique environmental system of particular importance to all coastal States. The draft envisioned the establishment of a regional organization of Caspian States as an institutional framework for future activities. Though the draft did not directly address the question of the legal status of the sea, its partition among coastal States, as well as any unilaterally asserted territorial claims, were implicitly rejected.

After a round of bilateral consultations only a few questions of minor importance remained unresolved before the final meeting of the Caspian States which was held in Moscow in October 1994. However, it became obvious at the meeting that Azerbaijan had reconsidered its position with respect to the Treaty, motivated primarily by the signing of its oil contract with a consortium of Western oil companies [40]. The Azeri delegation took the view that the conclusion of the Treaty and creation of a regional organization were premature. Particularly unacceptable for Azerbaijan were those provisions of the draft which treated the Caspian Sea as an area of joint utilization and called for coordination of approaches to the various activities related to the sea.

Since 1994 the initial tone of cooperation in relations among the Caspian States has radically changed. The current stance of the coastal States regarding the whole spectrum of the Caspian-related problems is determined by a number of factors, including these countries' relative dependence on energy resources as a source of foreign currency revenue, each State's geographical position with respect to the Caspian Sea and the location of the offshore petroleum deposits. For instance, whereas Russia, Iran and Kazakstan are already important producers of oil and have substantial deposits on land, Azerbaijan relies almost exclusively on the offshore resources. The most significant petroleum deposits are located in the southwestern and northeastern parts of the Caspian, closer to the coasts of Azerbaijan and Kazakstan, prompting these countries to press for division of the sea.

Three different approaches to the issue of the legal status of the Caspian Sea can be discerned from a survey of the relevant national legislation, adopted documents, draft agreements prepared by coastal States and their negotiating positions.

Azerbaijan regards the Caspian Sea as an "international or boundary lake" to be divided among the Caspian States in accordance with the sectoral approach [19]. To assert its claim over significant maritime areas, Azerbaijan in its recently adopted Constitution unilaterally declared sovereignty over its so-called "sector" of the Caspian Sea [5].

Kazakstan takes a less radical approach. It treats the Caspian as an enclosed sea subject to the principles and rules of the law of the sea, including those relating to the establishment of a territorial sea, exclusive economic zone and continental shelf by each coastal state. (The Draft Convention on the Legal Status of the Caspian Sea was prepared by Kazakstan in August 1994; the most recent version of the draft is dated from 20 October 1995.) It is evident, however, that given the size of the Caspian, this approach can only result in the actual partition of the sea among the littoral States [23].

Any division of the Caspian Sea is opposed by Russia and Iran, who contest all unilateral assertions of sovereignty and call for an approach based on joint management. The Russian position with respect to the legal status of the Caspian Sea and its natural resources is spelled out in a letter to the UN Secretary-General [32]. The joint development approach, however, does not go as far as to encompass the entire sea. Both Russia and Iran admit that the coastal States should be entitled to jurisdiction over their adjacent maritime areas within certain limits agreed upon by them. This approach was reflected in the working documents for negotiations prepared by Russia, one relating to the legal status of the Caspian Sea, and the other concerning exploration and exploitation of the mineral resources of the Caspian seabed (on file with the author). (On the Iranian position regarding the Caspian Sea issues, see [24].)

The Turkmenian position with respect to the Caspian is the most ambiguous and has been changed several times over the last few years. Turkmenistan was in fact the first littoral State to enact a law, in 1993, establishing territorial waters and an exclusive economic zone in the Caspian Sea. Later, it joined Russia and Iran in promoting the joint management of the Caspian and opposing unilateral development of its petroleum resources. This position is reflected in a paper by Turkmenian authors [26]. However, in July 1996, the Government of Turkmenistan announced its intention to hold a tender for the right to develop offshore oil and gas fields in its adjacent seabed areas [3]. This recent decision could indicate a reversal trend toward unilaterality in the Turkmenian position.

The Existing Legal Regime

The current legal regime of the Caspian Sea is based primarily on two agreements between the former Soviet Union (Soviet Russia) and Iran (Persia). The first — the Treaty of Friendship of 26 February 1921 — deals predominantly with political issues and applies the principle of equality of rights as the basis for bilateral relations between the two States [16]. However, only Article 11 of the Treaty, which established Iran's equal right of navigation, is of any relevance to the question of the legal status of the sea. Apart from this, the 1921 Treaty does not specifically address the issue of the legal regime of the Caspian Sea, as both countries were preoccupied mostly with their political relations in the post-World War I period.

The second — the Treaty of Commerce and Navigation of 25 March 1940 [30] — replaced the earlier Treaty of Establishment, Commerce and Navigation between the USSR and Iran of 27 August 1935 [16]. (The 1935 Treaty replaced the Convention of Establishment, Commerce and Navigation (October 27, 1931) between the USSR and Iran. The Convention reserved navigation and fishing rights in the Caspian Sea to vessels flying the Soviet or Iranian flag [Articles 16-17]). The principle of freedom of commerce and navigation for the nationals and ships of coastal States was reiterated, and an exclusive right to fish in their own coastal waters to a limit of 10 nautical miles was reserved to the vessels of each contracting party.

The underlying principle of both documents, which governs the utilization of the sea and its resources to the present date, is the exclusivity of rights of coastal States. The Caspian Sea is "closed" to all third States and their nationals. This situation has never been challenged by other States and was acknowledged by international legal doctrine [28, 12].

However, an exact definition of the legal status of the Caspian Sea has never been set out in any existing international instrument. This leaves such important issues as exploitation of the sea's mineral resources, protection of the marine environment or conservation and management of fisheries still to be addressed.

The Caspian has never been delimited between the littoral States and remained in their common and exclusive possession. Interestingly, in the Exchange of Notes that accompanied bilateral agreements between USSR and Iran, the Caspian Sea was repeatedly referred to as "the Soviet and Iranian (Persian) Sea." (See, for example, an Exchange of Notes between USSR and Persia of 1 October 1927 which emphasized a Parties' "mutual interest that the Caspian Sea remain exclusively Soviet-Persian" [16,17].) While this wording can be regarded as a political declaration reflecting the special interests of the coastal States, some authors consider it as a sufficient basis to define the legal status of the sea as a condominium. Dabiri, noting that "the Caspian Sea has always had a *sui generis* legal status," practically equates it to a condominium: "the letter and spirit of agreements reached between Iran and the former Soviet Union legally specify the Caspian Sea as condominium" [15]. However, no substantial evidence for such a claim can be discerned from the relevant practice of the USSR and Iran. There is no indication that the two coastal States have ever regarded the Sea as an area of their joint or shared sovereignty. (For a detailed analysis of the issue of condominium with respect to the Caspain Sea, see [11].) Rather, both States considered the Caspian as their *res communis*, exploited jointly by them but being beyond their territorial sovereignty. There were practically no limitations, except political and economic considerations or technological restraints, on the freedom to use the Sea and its resources by the two coastal States.

In many respects this *laissez faire* regime is comparable to that of the high seas before the 1958 Geneva Conference on the Law of the Sea. Each State, for example, was engaged in various activities in the Caspian without the prior consent of, or even consultations with, the other. One could argue that had the status quo in the region remained and the dissolution of the USSR not occurred, the Soviet Union and Iran could each have claimed rights with respect to their respective "parts" of the sea. However, any attempt to construct a legal argument based solely on a hypothetical situation is clearly not acceptable.

The sea remained their common resource outside any claims of sovereignty. The legal regime, established by the Soviet-Iranian treaties, no matter how outdated or incomplete, is still effective not only for Russia and Iran but also for the new coastal States, which explicitly agreed to be bound by the international obligations of the former Soviet Union [2].

The absence of a definitive conventional determination of the legal status of the Caspian leaves this issue open to argument and conflicting interpretation. Usually, the threshold question which is traditionally raised with respect to the Caspian Sea is whether this body of water is a sea or a lake? Evidently, the rules of international law applicable to each of these regimes are different. As a sea, the Caspian would be subject to the law of the sea rules embodied in large part in the 1982 United Nations Convention on the Law of the Sea (UNCLOS) [36]. If the Caspian were to be determined a lake, then the rules relating to the navigational and non-navigational uses of international watercourses would apply. In this respect, the recent work of the International Law Commission on international watercourses might indicate some of the rules that would govern certain uses of the Caspian [33].

For those familiar with the law of the sea there can be no doubt that the Caspian is not a sea *stricto sensu*, i.e., a maritime area constituting a part of the world ocean. It is not connected by an outlet to another sea or the ocean, and there are no territorial seas and exclusive economic zones of the coastal States in the Caspian. Thus, it fails to meet the definition of an enclosed or semi-enclosed sea under Article 122 of UNCLOS [40]. Article 122 of 1982 UNCLOS defines "enclosed or semi-enclosed sea" as "a gulf, basin, or sea surrounded by two or more States and connected to another sea or the ocean by a narrow outlet or consisting entirely or primarily of the territorial seas and exclusive economic zones of two or more coastal States" [36]. Obviously, the international law of the sea does not automatically apply to this body of water.

The Caspian Sea must be regarded as an international lake, although its size and geophysical features distinguish it from a typical lake. The Caspian is five times larger than the world's second largest lake, Lake Superior, and bigger than the Persian Gulf and the Oman Sea together [15, at 33]. Some experts qualify the Caspian as a relict marine basin (see [31]). The rules of the law of international watercourses might apply to certain uses of the Caspian waters and to protection of its environment. However, international law does not contain any steadfast rules governing uses of international lakes. Although in many cases international lakes have been delimited by littoral States, there are important exceptions to this practice [22]. Contrary to the opinion expressed by some writers [6], international law does not oblige coastal States to divide transboundary lakes, leaving this and other issues of their utilization at the discretion of States concerned.

It has become commonplace that the existing legal regime of the Caspian does not provide an adequate framework for addressing the new problems facing the region. New regulatory and institutional arrangements are required to deal with the new challenges and threats. In present circumstances, due to divergent and often contradictory views of the coastal States on the future legal status of the sea and particularly of its mineral resources, the comprehensive legal regime of the Caspian is hardly feasible. In certain areas, however, the concerted actions of the Caspian States are not only desirable but also possible without compromising their principal positions with respect to the issue of the legal status.

Areas of Possible Cooperation

SEA LEVEL RISE

Sea level rise is probably the most acute environmental challenge confronting the coastal States at present. This problem has already reached the dimensions of a major economic and social crisis affecting communities around the Caspian Sea, especially in the northern part where the topographic gradient is low. Historically, the level of the Sea has varied from year to year but has averaged at about 28 m (92 ft) below mean sea level. In the 1960s and 1970s the level fell substantially. Later, at the end of 1970s, the level of the sea started to increase. Since 1977 the water level has risen by more than 2 meters from –29 m to –27 m (below the global mean sea level), and according to some estimates this process will continue until at least 2005.

There is significant uncertainty as to the nature of this phenomenon which remains a matter of extensive scientific speculation, although climatic variables are considered to be the most probable cause. Because of insufficient knowledge and understanding of future climate behavior, it is impossible to predict precisely the time span and degree of sea level rise. However, the possible consequences of this process are easy to forecast. The rise in sea level to –25 m will result in the flooding and erosion of the Caspian coasts, including inundation of more than one million hectares of land. Several hundred thousand people would have to be removed from the regions in danger [6]. Flooding of the onshore hydrocarbon production areas is particularly threatening, as it will significantly increase contamination of the sea by oil and other oil-related harmful substances.

The pressing issue of sea level rise confronting the Caspian States urgently requires concerted measures on their part in order to remedy existing and future environmental threats. Adaptive strategies and policies, including rigorous integrated coastal zone management plans, laws and regulations, are to be adopted and implemented at both national and regional levels. Practical actions have to be preceded by coordinated scientific research efforts, including the exchange of information and joint studies of the climate-related environmental processes and problems.

Initially, all this can be done using existing national research centers and institutions. Later, some regional mechanism should be established with a mandate to coordinate research activities of cooperating coastal States. Eventually, the latter could be incorporated in a future regional organization, once a political decision to create such an organization has been reached.

Protection and Conservation of Living Marine Resources

Caspian living marine resources, primarily fisheries, have seriously declined over the last few years. Disintegration of the Soviet Union led to the collapse of the existing system of fisheries management and control, and to growing competition over limited fishery resources. Uncoordinated fishing efforts and, in particular, poaching, threaten the sustainability of Caspian fish stocks.

In the former Soviet Union, significant attempts were made to increase the sturgeon population through fish farming, two-thirds of which was concentrated within the Russian part of the Caspian coast. Additionally, at the beginning of 1960s, sturgeon fishing in the sea was banned and moved to the inflowing rivers and estuaries, thus targeting mature fish. A number of conservation measures allowed for the substantial reduction of the catch of undersized fish and to ensure sustainable harvesting of the sturgeon stock. However, this practice was recently abandoned, and fishing in the sea was resumed. Overfishing, poaching and water pollution have led to an almost fourfold decrease in the total catch of sturgeon [29].

The limited nature of fishery resources of the Caspian and their vulnerability to over-exploitation requires coordination of coastal States' fishing policies and activities. Regional management of living marine resources has been introduced and has proven to be effective in various parts of the world. With this in view the coastal States undertook to draft a special agreement regarding biological resources of the Caspian Sea. After a series of bilateral and multilateral consultations, a draft Agreement on the Conservation and Utilization of the Biological Resources of the Caspian Sea was prepared [18]. A meeting of Caspian States was convened in 30 January - 2 February 1995, in Ashgabat (Turkmenistan), in order to finalize the draft. The only major issue which remained unresolved before the meeting was the future extent of the exclusive fishing jurisdiction of coastal States. Positions of the parties varied from 15 (Russia) and 25 (Kazakstan) nautical miles to 30 (Iran) and 40 (Azerbaijan and Turkmenistan) nautical miles. In the end, four Caspian States (Iran, Kazakstan, Russia and Turkmenistan) agreed on the 20-mile limit of the coastal States' fishing zones and adopted the final text of the Agreement. Azerbaijan refused to accept the text on the premise that the adoption of the Agreement would predetermine the final decision on the legal status of the Caspian [27].

The draft Agreement views the Caspian Sea as a unitary ecological system [18]. The biological resources of the Caspian Sea are regarded as common to all coastal States, and the latter bear special responsibility for their conservation, reproduction and optimal utilization [18]. Under the draft, only the nationals and/or juridical persons of the Caspian States have a right to fish in the treaty area [18]. (Under Article 1, the area encompasses the whole of the Caspian Sea including parts of the rivers, where migratory species of fish spawn or migrate.) Coastal States can establish zones of fishery jurisdiction up to 20 nautical miles from their coasts, provided that fishing vessels of other States will be allowed to continue fishing in these zones on the basis of licenses [18, Article 4(1) and (2).] In order not to prejudice the final decision regarding the legal status of the sea, two safeguard clauses were specifically envisaged. The first provides that the breadth of fishing zones are susceptible to changes depending on the future agreement regarding the legal status of the Caspian Sea. According to the second, establishment of such zones does not affect exploitation of mineral resources, navigation and other activities not related to the utilization of biological resources.

The draft provides for the use of quotas as a principal instrument of fishery management. Fishing of sturgeon in the open sea is explicitly prohibited and is limited to the inflowing rivers and river estuaries within the territories of coastal States. (A special amendment to the draft Agreement provides for the only exception to this rule. Iran was allowed to continue traditional fishing of sturgeon near its coast within the limits of its quota [18].) The access of fishermen of other Caspian States is conditioned and based on the quotas allocated to respective coastal States. Under the draft, the parties are expected also to coordinate their export policies, prices and export quotas with respect to fish products and sturgeon caviar.

The draft Agreement envisages the creation of a special institutional mechanism — a Committee on the Conservation and Utilization of the Biological Resources of the Caspian Sea — with a fairly broad mandate. It is empowered with obligatory decision-making regarding a wide range of issues, including: determination of the total allowable catch of fish stocks, as well as other living marine resources, in the area beyond the limits of national jurisdiction; allocation of quotas among the parties; coordination of fishing and conservation policies; adoption of fishery regulatory and conservation measures [18, Articles 10 and 11].

Despite some flaws and inconsistencies, the draft Agreement is a balanced document, which adequately reflects both the interests of coastal States and the requirement of rational utilization and conservation of the Caspian living marine resources. Given its special provisions, which take into account possible changes in the legal status of the sea, the opposition of Azerbaijan to the Agreement seems to be unfounded. However, it can be understood in view of Azerbaijan's general reluctance to accept anything which might undermine its drive for a complete division of the Caspian. Unfortunately, this opposition to the Agreement precludes, at least for the time being, adoption of regulatory and conservation measures, the urgency of which is obvious.

Protection of the Marine Environment

The Caspian Sea represents a unique and extremely vulnerable ecosystem. Without a direct connection with the ocean, it has a very low absorbing and self-cleaning capacity and is particularly susceptible to potential adverse impacts of land-based and offshore economic activities.

Environmental protection of the Caspian has never been a special subject of Soviet-Iranian arrangements, but rather has been dealt with at the national level [8, 9]. The only relevant provision of the 1940 Treaty is Article 15 which reads:

> The high contracting Parties are agreed that, in respect of sanitary measures to be undertaken in the case of the vessels of either of the high contracting Parties in the Caspian ports of the other, the provisions of the International Sanitary Convention signed in Paris on June 21, 1926, with due regard to the observations made by each of the high contracting Parties at the time of the signature, shall be applied [30].

At present, protection of the marine environment against pollution resulting from offshore development, particularly in cases of possible accidents and oil spills, is a matter of primary concern. Another issue, which should be addressed within the context of environmental cooperation, is protection of those areas of the Caspian which are particularly important for the reproduction of its living marine resources [18, Article 4(3)]. In fact, in the 1970s the governments of the former Russian and Kazak Soviet Republics established specially protected areas in the northern part of the Caspian Sea and the deltas of the Volga and Ural rivers for the purpose of conservation and reproduction of fishery resources [20]. Under these regulations any form of economic activities in the designated areas, except those related to fisheries and navigation, were completely prohibited. The ban applied to the exploration of the seabed, including seismic activities, drilling of wells and their exploitation. Practically the whole northern part of the Sea was excluded from economic development.

Ironically, this specially protected area of the Caspian off the coast of Kazakstan has turned, during the last few years, into a region of intensive petroleum exploration activities conducted by Kazakhstankaspishelf, a consortium which includes British Gas, BP, Agip, Mobil, Shell and other oil companies. According to approximate estimates of

Kazak authorities, which are based on the preliminary results of the seismic studies of this part of the sea, its potential reserves of crude oil may constitute 10 billion metric tons and of natural gas 2 trillion cubic meters [4]. If confirmed, these offshore oil reserves would be ten times larger than those of its onshore Tengiz oil field. Plans to begin offshore drilling are presently under way in this and other parts of the sea. The hydrocarbon development on a large scale has already started off the coast of Azerbaijan. (For a detailed survey of the Caspian petroleum potential and analysis of related legal problems, see [38].)

Offshore petroleum exploration, exploitation and transportation are by their very nature hazardous activities associated with extreme risk for the marine environment. This is particularly true with respect to the entirely enclosed Caspian Sea, where any significant oil spill, the possibility of which cannot be absolutely excluded, can cause irremedial harm. In these circumstances preventive and precautionary measures have to be introduced and implemented on both national and international levels. (Principles of prevention and precaution have been widely accepted and included into all recent international instruments dealing with protection of the marine environment. See [13, 14].) No individual State or oil company is able to cope with these risks alone. A regional network of environmental monitoring, emergency response and assistance must precede any large-scale offshore petroleum development.

Ideally, an international legal framework prescribing general obligations and addressing specific environmental problems is required, as well as institutional mechanisms for environmental cooperation at the regional level. To avoid the risk of prejudice to the coastal States' positions with respect to other aspects of the Caspian legal regime, a special disclaimer clause could be used. An example of such a provision can be found in the 1976 Barcelona Convention on the Protection of the Mediterranean Sea Against Pollution. Article 3 (1) stipulates that:

> Nothing in this Convention shall prejudice ... the present and future claims and legal views of any State concerning the law of the sea and the nature and extent of coastal and flag State jurisdiction [14].

In the preparation of such an instrument the available State practice and experience of international organizations, primarily that of UNEP, might be taken into account [25]. In this respect a two-step approach, adopted by UNEP, seems well suited for the Caspian. A "framework" convention, containing general legal obligations, can be supplemented by a set of protocols addressing specific issues. Of primary concern among those are protection of areas particularly important for the conservation of living marine resources, prevention of pollution from land-based sources and from offshore exploitation of hydrocarbons and their transport, cooperation in combatting pollution in cases of emergency and in dealing with the detrimental consequences of sea level rise.

The Caspian States have been urged recently to take concerted actions to ensure protection of the environment and sustainable utilization of natural resources in the region. Representatives of all coastal States, gathered by UNEP in Geneva, on 12-14 December 1995, acknowledged the deteriorating ecological situation in the Caspian region [34]. A series of actions for a regional environmental program were proposed, based on recommendations of the joint UNDP/UNEP/World Bank mission to the Caspian region, which took place in March-April 1995.

UNEP was requested to act as a facilitator in the development of a framework convention for the Caspian region. It was agreed that this process should involve all Caspian States and should be without prejudice to the issues of the legal status of the Caspian. The need for further improvement and harmonization of national legislation and institutions was also recognized. To ensure UNEP's facilitating role, the experts recommended that the Caspian States initiate the adoption of a relevant decision by the 19th session of the Governing Council in January 1997 [35].

Concluding Remarks

The Caspian Sea, although geographically and legally not a part of the world ocean, can yet qualify as a large marine ecosystem, a relatively new notion of a distinct entity which requires a holistic approach to its management [1]. Ideally, utilization and conservation of all natural resources of the Caspian should be dealt with in a comprehensive manner. In theory, the Caspian Sea could become a perfect model in applying the concepts of sustainable development and integrated management of its ecosystem and natural resources as a unitary whole.

In reality, the divergent political and economic aspirations of the coastal States together with the direct involvement of outside players undermine potentially advantageous regional cooperation in its most advanced form. Conspicuous focus of some States on the question of petroleum resources only inhibits even modest progress towards any kind of joint actions.

However, there are problems which need to be dealt with in the immediate future. Conservation of living marine resources, protection of the marine environment, along with the issues of the sea level rise, are the areas which are ready and call for the concerted actions of all coastal States. Although controversy regarding the status of the sea and its resources, and especially unilateral claims and actions on the part of individual coastal States, can hardly create favorable conditions for required cooperation, there is no option. Environmental threats facing the region are real and any delay with the necessary measures can only make the situation worse.

Acknowledgments

The author would like to thank Patricia Wouters for her comments and contribution to this chapter. Viewed expressed in this work are the author's alone and do not necessarily reflect the position of the Centre for Petroleum and Mineral Law and Policy.

References

1. Alexander, L., 1993: Large marine ecosystems: A new focus for marine resources management. 17 *Marine Policy*, 186-198.

2. Alma Ata Declaration, 1992. *I.L.M.*, **31**, 149. The Declaration, which was signed by the former Soviet republics, including three new Caspian States, provides: "The States participants in the Commonwealth guarantee in accordance with their constitutional procedures the discharge of the international obligations deriving from treaties and agreements ocncluded by the former Union of Soviet Social Republics," Alma Ata Declaration 21 December 1991.

3. Anon., 1996: Expanding horizons: Turkmenistan tender for Caspian reserves marks shift from Russia toward the West. *Russian Petroleum Investor*, **52** (September).

4. Anon., 1996: Kazakhstan reports huge offshore oil reserves. *Open Media Research Institute Daily Digest*, No. 125, Part 1, 27 June 1996. Internet.

5. Article 12 of the new Constitution of Azerbaijan proclaims the so-called Azeri sector of the Caspian Sea as a part of its national territory, in: Constitution of the Republic of Azerbaijan [document on file with author].

6. Carver, J., and G. Englefield, 1995: The future development of the Caspian Sea. *Oil and Gas Law and Taxation Review*, **1**, 43. According to Carver and Englefield, "If the Caspian Sea were treated as an inland lake, the littoral States would have to divide the whole area between them, each sector being under the absolute sovereignty of the relevant coastal state with no rights of innocent passage."

7. Communiqué of the Conference on the Problems of the Caspian Sea, 14 October 1993. Astrakhan [translated by author; on file with author].

8. Council of Ministers of the USSR, 1977: *On Additional Measures Relating to the Protection of the Caspian Sea Against Pollution*, No. 990, 16 November 1977. Moscow, Russia: Council of Ministers.

9. Council of Ministers of the USSR, 1968: *On Measures Relating to the Prevention of Pollution of the Caspian Sea*, No. 759, 23 September 1968. Moscow, Russia: Council of Ministers. Contains a wide range of measures to eliminate and control various sources of pollution.

10. British and Foreign State Papers, 1968: Vol. 134, 1026. London, UK: Foreign Office.

11. Bundy, R., 1995: The Caspian: Sea or lake? Consequences in international law. In C. Gurdon and S. Lloyd (eds.), *Oil and Caviar in the Caspian*, 16-21. London, UK.

12. Butler, W.E., 1969: The Soviet Union and the Continental Shelf. Notes and Comments, *American Journal of International Law,* **63**, 106. Butler acknowledges this fact: "Soviet jurists regard the Caspian as a large lake which historically has been called a sea. General norms of international law relative to the high seas, to vessels and their crews sailing on the high seas do not extend to the Caspian, whose regime is governed by Soviet-Iranian treaties and agreements."

13. Convention for the Protection of the Marine Environment of the Northeast Atlantic, 1993: Paris, France, 22 September 1992. *I.L.M,*. **32**, 1069.

14. Convention for the Protection of the Mediterranean Sea Against Pollution, 1976: Barcelona, Spain, 16 February 1976. *I.L.M.*, **15**, 290.

15. Dabiri, M.-R., 1994: A new approach to the legal regime of the Caspian Sea as a basis for peace and development, *Iranian Journal of International Affairs*, 28-46 (Spring/Summer).

16. Documenty vneshney politiki SSSR (Documents of the Foreign Policy of the USSR), 1968. Moscow, Russia, 603-604.

17. Documenty vneshney politiki SSSR (Documents of the Foreign Policy of the USSR), 1959. Moscow, Russia, 536.

18. Draft Agreement on the Conservation and Utilization of the Biological Resources of the Caspian Sea [on file with author].

19. Draft Convention on the Legal Status of the Caspian Sea, prepared by Azerbaijan, 12 October 1995 [on file with author].

20. Enactment of the Council of Ministers of the RSFSR "On the Proclamation of the Specially Protected Area in the Northern Part of the Caspian Sea," No. 78, 31 January 1975 (in Russian); Enactment of the Council of Ministers of SSR of Kazakhstan "On the Proclamation of the Specially Protected Area in the Northern Part of the Caspian Sea," No. 252, 30 April 1974 (in Russian).

21. Englefield, G., 1995: A spider's web: Jurisdictional problems in the Caspian Sea. *Boundary and Security Bulletin,* **3**, 30-33.

22. Frowein, J.A., 1989: Lake Constance. In R. Bernhardt (ed.), *Encyclopaedia of Public International Law*, Installment 12, at 216.

23. Gizzatov, V., 1996: A legal status of the Caspian Sea: Condominium or delimitation. Paper presented at conference, *The Caspian Sea: Political Realignments, Energy Implications and New Security Architectore for the Twenty-First Century*, Sandhurst Royal Military Academy, 13-14 May 1996. Sandhurst, UK: Conflict Studies Research Centre.

24. Granmayeh, A., 1995: The Caspian Sea in Iranian history and politics. *Labyrinthi Central Asia Quarterly*, **2**, 36-40 (Summer).

25. Haas, P., 1991: Save the seas: UNEP's Regional Seas Program and the coordination of regional pollution control efforts. *Ocean Yearbook*, **9**, 188-212.

26. Kochumov, J.Kh., and E.A. Kepbanov, 1996: Novyi pravovoi status Kaspiia -- osnova obespecheniia mira i stabilnosti v regione (New legal status of the Caspian: A basis for peace and stability in the region), presented at conference, *The Caspian Sea: Political Realignments, Energy Implications and New Security Architecture for the Twenty-First Century*, Sandhurst Royal Military Academy, 13-14 May 1996. Sandhurst, UK: Conflict Studies Research Centre.

27. Minutes of the negotiations of the Caspian States Delegations concerning Draft Agreement on the Conservation and Utilization of the Biological Resources of the Caspian Sea, 30 January to 2 February 1995, Attachment 3, Separate opinion of the Delegation of the Republic of Azerbaijan [on file with author].

28. Pondaven, Ph., 1972: Les lacs-frontière, Paris, France.

29. *Rossiyskaya Gazeta*, 11 November 1994, at 5. Whereas in the middle of the 1970s the total catch of sturgeon in the USSR had reached 27,000 tonnes, in 1994 all four former Soviet republics produced only 4,000 tonnes plus 3,000 tonnes were caught by Iran.

30. Sbornik deistvouyuschikh dogovorov, soglasheniy i konventsiy, zaklyuchyonnikh SSSR s inostrannymi gosoudarstvami (Collection of treaties, agreements and conventions concluded by the USSR with foreign States), 1955: **10**, 56. Moscow, Russia.

31. United Nations, 1995: *Minutes of the Meeting on Cooperation of UN Organizations in the Caspian Sea Initiative*, 17 January. Geneva, Switzerland: United Nations.

32. United Nations, 1994: Position of the Russian Federation regarding the legal regime of the Caspian Sea. UN Doc. A/49/475, 5 October. The document states, *inter alia*, that "The Caspian Sea and its resources are of vital importance to all the States bordering on it. For this reason, all utilization of the Caspian Sea, in particular the development of the mineral resources of the Caspian seabed and the rational use of its living resources ... must be the subject of concerted action on the part of all States bordering the Caspian."

33. United Nations, 1994: *Draft Articles and Commentaries on the Law of the Non-Navigational Uses of International Watercourses. Adopted on Second Reading by the ILC at its Forty-Sixth Session*, UN Doc. A/CN.4./L.493 and Add. 1 and Add.2 (12 July 1994).

34. UNEP (United Nations Environment Programme), 1996: The Caspian region: A leading role for UNEP. *UNEP Update, No. 1*, February, Internet.

35. UNEP, 1995: Meeting of Experts on Environmental Cooperation in the Caspian Region: Legal Framework and Institutional Mechanisms. UNEP-ROE/CASP.LG1/1. Geneva, Switzerland: UNEP.

36. United Nations, 1983: *The Law of the Sea. UN Convention on the Law of the Sea with Index and Final Text of the Third United Nations Conference on the Law of the Sea*. New York: United Nations.

37. Vallega, A., 1994: The regional scale of ocean management and marine region building, 24 *Ocean and Coastal Management*, 17-37.

38. Vinogradov, S., 1996: The legal status of the Caspian Sea and its hydrocarbon resources. Paper presented at *Fourth International Conference of the International Boundaries Research Unit (IBRU), Boundaries and Energy: Problems and Prospects*, 18-19 July 1996, University of Durham.

39. Vinogradov, S., and P. Wouters, 1996: The Caspian Sea: Quest for a new legal regime, *Leiden Journal of International Law,* **9**, 87-98.

40. Vinogradov, S., and P. Wouters, 1995: The Caspian Sea: Current legal problems. *55 Zeitschrift für ausländisches öffentliches Recht und Völkerrecht,* 604-623.

II. Sea Level Rise: Impacts and Implications

IRANIAN RESEARCH ON THE CASPIAN SEA LEVEL RISE

JALAL SHAYEGAN
Sharif University of Technology
Tehran, Iran

AMIR BADAKHSHAN
University of Calgary
Calgary, Alberta, Canada

The level of the Caspian Sea has risen over 2.5 meters in nearly 20 years and is continuing to rise. Fluctuations in the level of this sea have been observed in the past, but its present rate of increase is without precedent.

In this chapter the probable causes of such a high rate of increase in the sea level are identified and explored. The adverse economic impacts of such an occurrence on agricultural, human settlements, recreational, and commercial activities along the Iranian shore are noted, as are ways to combat the problem.

Introduction

The Caspian Sea is the largest inland lake in the world. It is a remnant of a very large ancient sea (Tethys) which had once covered the total area of Iran, Turkey, Southeast of Russia and the Mediterranean Sea. The Caspian Sea is divided into northern, middle and southern sections.

The length of the Caspian Sea from south to north is 1,204 km, and its width ranges from 196 to 448 km. Its surface area is 429,140 km^2 with its volume of water at 78,000 km^3. The shoreline of the sea is 5,200 km in length; the Iranian shore portion is 724 km. The Caspian Sea takes in 260-340 km^3/year of water of which 5% is from Iranian rivers. The sea which is located on the north of Iran is surrounded on the east by Turkmenistan, on the east and north by Kazakstan, and on the west by Russia and Azerbaijan. The salinity of its water is 12.7 g/l as compared with 35 g/l in the world's oceans. The Kara-Bogaz-Gol Bay, on the Turkmenistan shore, is the largest gulf of the sea. This gulf plays an important role in the equilibrium of the water level in the sea. The Kara-Bogaz-Gol Bay is a location for the commercial extraction of various salts which are formed by the evaporation of the sea water.

M.H. Glantz and I.S. Zonn (eds.),
Scientific, Environmental, and Political Issues in the Circum-Caspian Region, 69-77.
© 1997 *Kluwer Academic Publishers. Printed in the Netherlands.*

Historic Trend of Caspian Sea Water Level Fluctuation

Historic observations and archaeological studies have revealed that the Caspian Sea level has shown variations of up to 8 meters in the past 2000 years. Beginning in 1930, the level declined. Its lowest value occurred in 1977 at −29 m. After 1977, the level of the sea started to increase at an alarming rate reaching a value of more than −27 m and is still increasing.

Causes of Water Level Fluctuation

Possible reasons for water level fluctuations are as follows:

1. Long-term, cyclical changes in level.

2. The use of sea water, dam construction and diversion of streamflow en route to the sea for agriculture and/or industrial activities.

3. Tectonic movement of the sea bottom and the creation of marine springs.

4. Climate changes, causing variations in inflow of river water into the sea and an increase in the level of precipitation.

5. Reduction of evaporation from the sea's surface because of oil pollution.

6. Diverting the routes of non-contributing rivers for delivery of their flow to the sea.

7. Control of water inflow rates to the sea and to the Kara-Bogaz-Gol Bay and control of evaporation from the sea.

8. Greenhouse effect and melting of permafrost.

9. Creation and destruction of water ways between the Black Sea and the Aral Sea with the Caspian Sea.

The reduction of the water level in the Caspian Sea after 1930 corresponds to the development of industrial and agricultural mega-projects in the former USSR. Changing the bed of certain rivers, constructing artificial lakes and dams and intensifying human activities in these areas could have been the reason for the lowest sea level (−29 m) in the year 1977. The adverse impacts of such a low water level on economics and the environment in the former USSR were profound. These adverse effects are summarized as follows:

1. A reduction of fishing activities.
2. An increase in water salinity.
3. Navigation problems because of the decreasing depth of water in channels and ports.
4. A reduction of aquifer levels.
5. Creation of desert conditions to the east of the Volga River.

The lowering of sea level along Iranian shores was considered to have been permanent and the newly exposed sea bottom was used for the construction of settlements, for the expansion of urban and agricultural activities and for animal husbandry.

The devastating effects on the life of the people because of the lowering of the Caspian Sea level, were realized in Russia in 1977. Since then, modification of existing water use plans and the creation of new plans were designed to correct the situation, resulting in a high rate of sea level rise.

Losses Encountered by the Rise of the Caspian Sea

Studies have shown that, although the level of the sea started to increase in 1978, its devastating effects did not occur until 1986. The losses caused by the impact of sea level rise were more noticeable when torrential rainy periods were combined with high winds. Although the detailed magnitude of losses was not provided, titles appeared in the news such as the following: "Devastating Impact of Sea Waves on Mazandran," "The Magnitude of Losses Created by Sea Floods and Advance of Water on the Shores of Caspian Sea Are Very Heavy," and "Astara on the Verge of Complete Flooding." Such headlines are good indicators of the magnitude of such disasters.

Losses can be classified in three categories:

1. Financial losses.
2. Land and agricultural losses.
3. Human lives and devastating psychological impacts on people.

A quantification of financial losses and of land and agricultural losses using data collected by Ministries of State and Power and from visits to the affected areas is not difficult. However, an assessment of psychological impacts and lives lost, no doubt most important, has been seriously neglected. This is an important item for research in the future.

Financial Losses

About 10,000 houses were damaged or destroyed. Infrastructure losses include damages to water and electric power installations, port facilities and fisheries institutions.

The financial equivalent of losses are estimated at 1500 billion Rials, of which 700 billion is only for the Province of Mazandran. Furthermore, the cost of relocation of destroyed houses and institutions and the construction of water barriers and wave breakers has been estimated at 30 billion Rials.

Land and Agricultural Losses

The shoreline of Iran on the Caspian Sea is nearly 750 km. The low-lying parts of the shore have so far permitted about a 22-meter water advance landward, which means the flooding of 20,000 hectares of land. This flood has caused damage to water, telephone and electricity networks, irrigation canals, office buildings, deep and semi-deep wells, orchards, agricultural lands, beaches and fishery stations, hotels and guest houses, piers and port facilities, storage facilities for agricultural products, bridges and communication networks, power stations and custom buildings.

Forecast of Future Losses

Experience, observation and data analyses have indicated that the Caspian Sea level will reach an equilibrium value at about −24 m. A sea level of −24 m means that a level of −23 m would be a logical and economic level for future installations for shore cities on the Iranian side. An acceptance of the −23 m level means the following:

1. At levels −23 m to −24 m on the shores of Guilan province (Astara, Anzali, Kiashahr, Roudsar, Kalachy, Chaboksar), 3154 hectares of land and 2319 houses would be flooded or require relocation and/or reconstruction.

2. At levels of −24 m or less, 1244 hectares of urban areas and 4900 houses would be threatened by rising sea water. The loss of rural area is calculated to be 80 hectares. Five billion Rials would be required for the undertaking of necessary "precautionary" expenditures.

In general, at a level of −23 m, along the Iranian shore, 20,000 hectares of land and about 4 million m^2 of residents, office buildings and infrastructure exist which would require about 1000 billion Rials for their relocation. The cost of construction of 500 km of sea walls has been estimated at about 1250 billion Rials.

Other Environmental Hazards Resulting From Sea Level Rise

1. The introduction of new sources of eutrophication such as nitrogen and phosphorous and also pesticides and herbicides left over from agricultural activities on farmland covered by the sea water.

2. The penetration of salt water into wetlands and drinking-water wells.

3. The entrainment of different kinds of municipal and industrial solid wastes as well as all type of vegetation (trees, grasses, etc.) to the sea as a result of the advancement of water.

4. The destruction of bird habitats and spawning areas for fish.

5. Coastal erosion and the introduction of large amounts of soil (sediments) into the sea, causing continuous changes in marine, coastal and land ecosystems.

Discussions and Conclusions

1. The rise of Caspian Sea level is a grave problem for Iran and requires keen attention at the national, regional and international levels.

2. Close collaboration is urgently required among countries surrounding the Caspian Sea in order to investigate the real parameters of such a phenomenon.

3. In-depth feasibility studies for combating the problem and optimization of possible choices, e.g., relocation vs. wall construction, are strongly recommended.

4. The environmental impact of the sea level problem and its possible relation to global warming should be seriously undertaken by national, regional and international authorities (e.g., UNEP, UNESCO, FAO, etc.).

5. If the rise of the Caspian Sea level has some advantages for the countries in the region, its magnitude should be assessed and the countries who are disadvantaged by the rise should be compensated.

6. Urgent attention to these issues will no doubt decrease the level of damage and would help to prevent avoidable disturbances to human settlements and to human activities in the future.

TABLE 1: Physical Characteristics - Caspian Sea

Sea Area		**Width**	
Total	429,140 km^2	Max.	448 km
Northern	126,596 km^2	Min.	196 km
Middle	151,626 km^2	Average	330 km
Southern	151,018 km^2	**Maximum Depth**	
Basin Area		Northern	26 m
Total	3.77×10^6 km^2	Middle	958 m
USSR (Old)	3.5×10^6 km^2	Southern	960 m
IRAN	264,000 km^2	**Periphery**	
Volume	78×10^3 km^3	Total	5,200 km
Length	1,204 km	Iranian Side	724 km

TABLE 2: Caspian Sea Surface Water inflow ($10^9 m^3$/Year)

Periods	Surface Water Input		Volga Share %
	Volga River	Total	
1900-1929	250.6	335.7	74.6
1930-1941	200.5	268.6	74.6
1942-1969	241.2	285.4	84.5
1970-1977	207.6	240.4	86.3
1978-1993	274.3	310.9	88.2

TABLE 3: Water Balance & Level Changes From 1900-1993 ($10^9 m^3$)

Periods	Total Inflow (Surface Underground & Precipitation)	Outflow			Water Volume Changes	
		Evaporation	Kara Bogaz	Total	Period	Year
1900-1929	409.4	389.4	21.8	411.2	-13.1	-0.5
1930-1941	347.1	394.8	12.4	407.2	-180	-15
1942-1969	369.9	356.3	10.6	366.9	+22	+0.8
1970-1977	342.2	374.9	7.1	382	-79.6	-10
1978-1993	418	355	2.5	357.5	+228	15.2

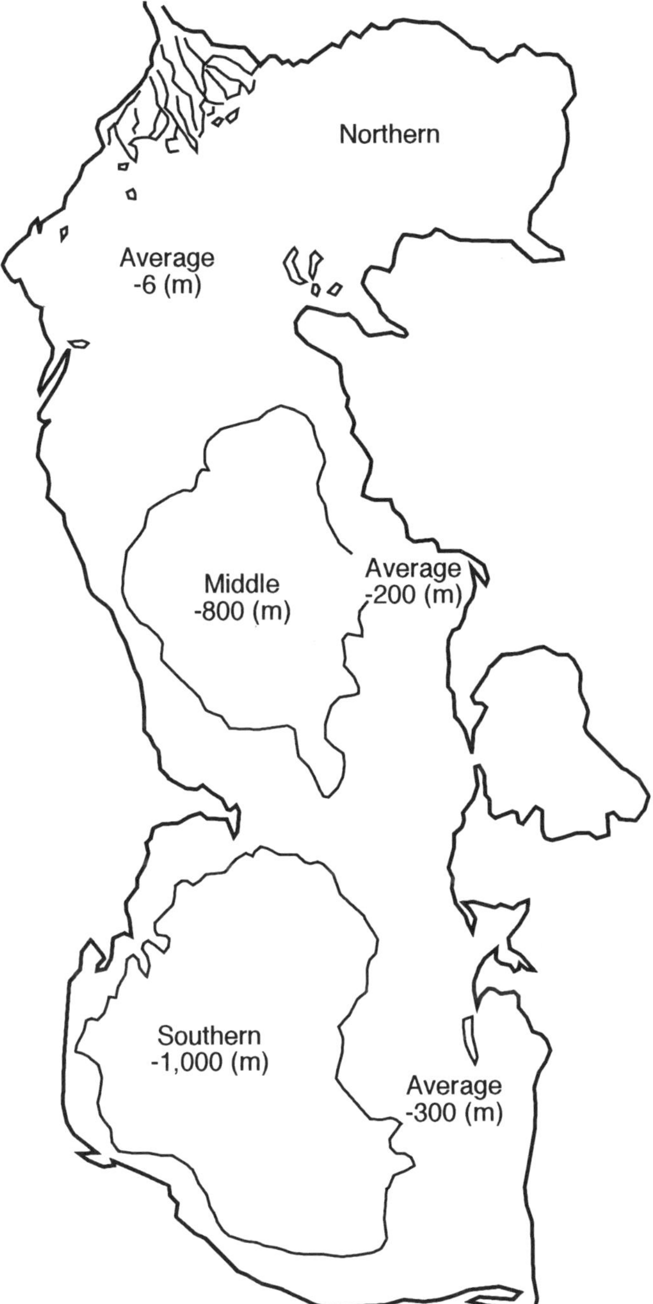

Fig. 1. Different sections of the Caspian Sea

Fig. 2. Currents of the Caspian Sea

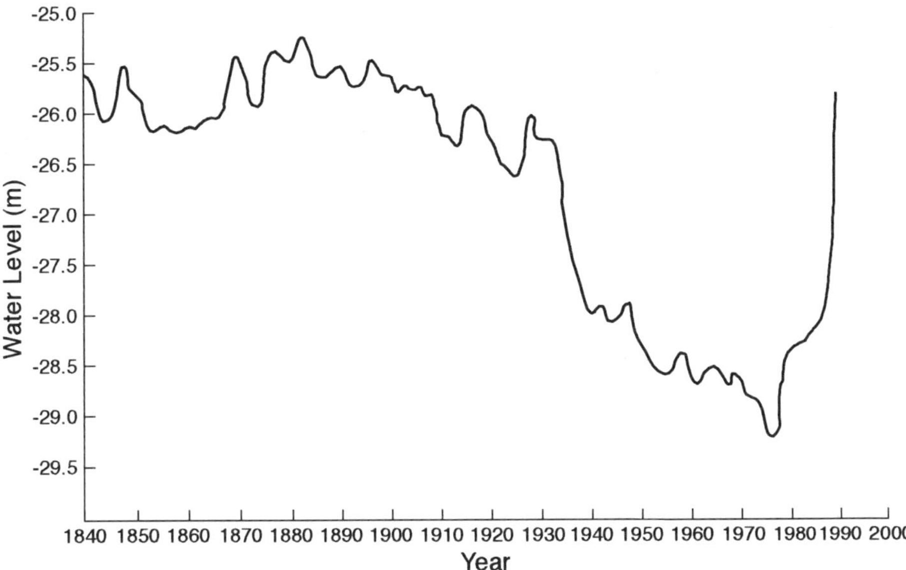

Fig. 3. Caspian Sea level changes 1840-1994

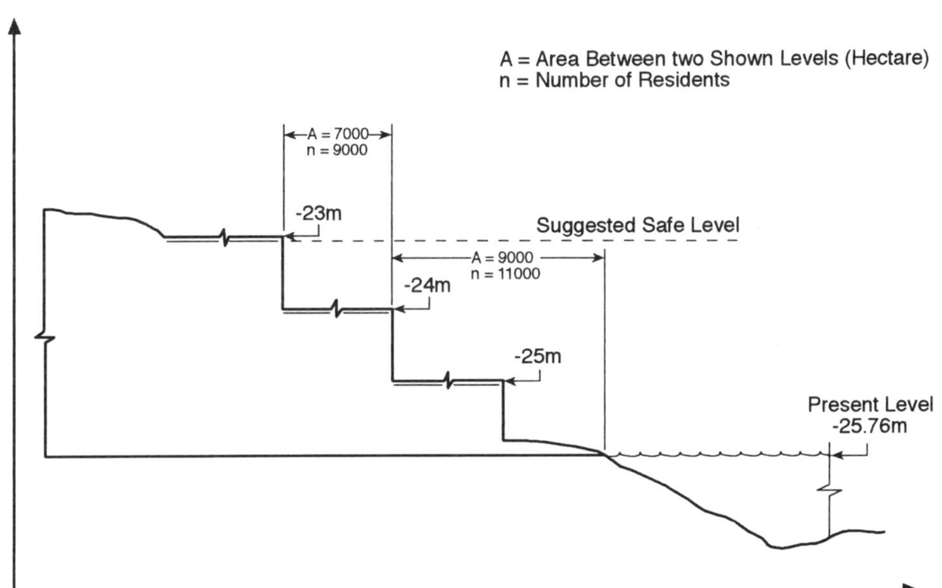

Fig. 4. Number of houses and areas of land in danger

ECOLOGICAL PROBLEMS OF THE CASPIAN SEA AND PERSPECTIVES ON POSSIBLE SOLUTIONS

ARIF SH. MEKHTIEV and A.K. GUL
Azerbaijan National Aerospace Agency
Baku, Azerbaijan

The Caspian Sea is 1,204 km from north to south and 325 km from west to east. The length of the coastline is 7,000 km. Its maximum depth is 1025 m and average depth is 207 m. The volume of water is about 78,000 km^3, or more than 40% of the total of lake water on earth [1]. A distinctive feature of the Caspian Sea is its rather significant periodic changes in level, causing different ecological and social problems around its coastline.

Caspian Sea Level Changes

In 1954, Aliyev [2] published interesting information about Caspian Sea level from the first century B.C. to the 10th century A.D. (Table 1):

TABLE 1. Caspian Sea level Changes

Period	Level (m)
1st Century B.C.	−31.5
63 B.C.	−27.5
2nd Century	−26.0
7th Century	−30.0
10th Century	−26.0

From the 10th to the 16th century, maps of the Caspian Sea showed the sea level to be about −30.5 m. In 1556 a level of −26.4 m. was noted. Berg [3] reported that for the last 400 years the levels rose and fell periodically. From the beginning of systematic observations of Caspian levels (in 1830), four characteristic periods can be identified: a relatively stable period 1830-1929 (−25.5 m); a sharp

M.H. Glantz and I.S. Zonn (eds.),
Scientific, Environmental, and Political Issues in the Circum-Caspian Region, 79-95.
© 1997 *Kluwer Academic Publishers. Printed in the Netherlands.*

1.7 m fall between 1930 and 1941; and then, beginning in 1942, there was a slow decrease until 1977. At the end of this period the Caspian Sea level was at its lowest point in the last 160 years at −29.0 m. The total decrease in level between 1900 and 1977 was 3 m. Since 1978 the level of the Caspian Sea has been increasing and up to the present it has increased almost 2.55 m. (Fig. 1).

The change in Caspian Sea level may have been the result of any one or combination of various causes: climate change, tectonic processes and human activity. The influence of each of these factors as a contributor of the total changes in level is not equal.

Fedorov [4], who studied the modern geologic record of the sea in detail, has established that long-term increases and decreases of the Caspian level were defined primarily by climatic factors. He suggested that tectonic movement in the Caspian region did not have a significant influence on the level.

Rychagov [5] agrees with this view. He believes that it is possible to explain the frequent and significant fluctuations of Caspian Sea level in the Holocene only by climate changes. In his opinion, tectonic shifts could not precipitate an appreciable change in level.

Among the reasons for changes of the Caspian Sea level, groundwater flow has been suggested. However, according to a 1984 study [6], the total amount of groundwater from along the entire perimeter of the sea (estimated at 3.2 km^3 per year) would account only for about 2-3% of the total sea level increase.

According to calculations by Kosarev [7], human activities have been an important factor affecting sea level. He showed that the volume of used but unreturned water in the 1970s amounted to 40-50 km^3 per year, causing a decrease in level of up to 10-12 cm annually. This suggests that, under natural conditions, the sea level could have been 1.5-2.0 m higher than the present level.

It is now generally accepted that the key factor affecting fluctuations of Caspian Sea level is climate in the huge Caspian Sea basin. The contribution of climate to sea level changes is about 85% according to Kosarev [7]. He shows in Table 2 that in all cases the increase in level was the result of river flow to the Caspian.

One can also see a reduction in evaporation, a main component in the water budget since 1978 compared to the norm (100 cm) and an increase in the annual amount of precipitation (22 cm) when compared to the annual average (19 cm) for a period beginning at the turn of the century.

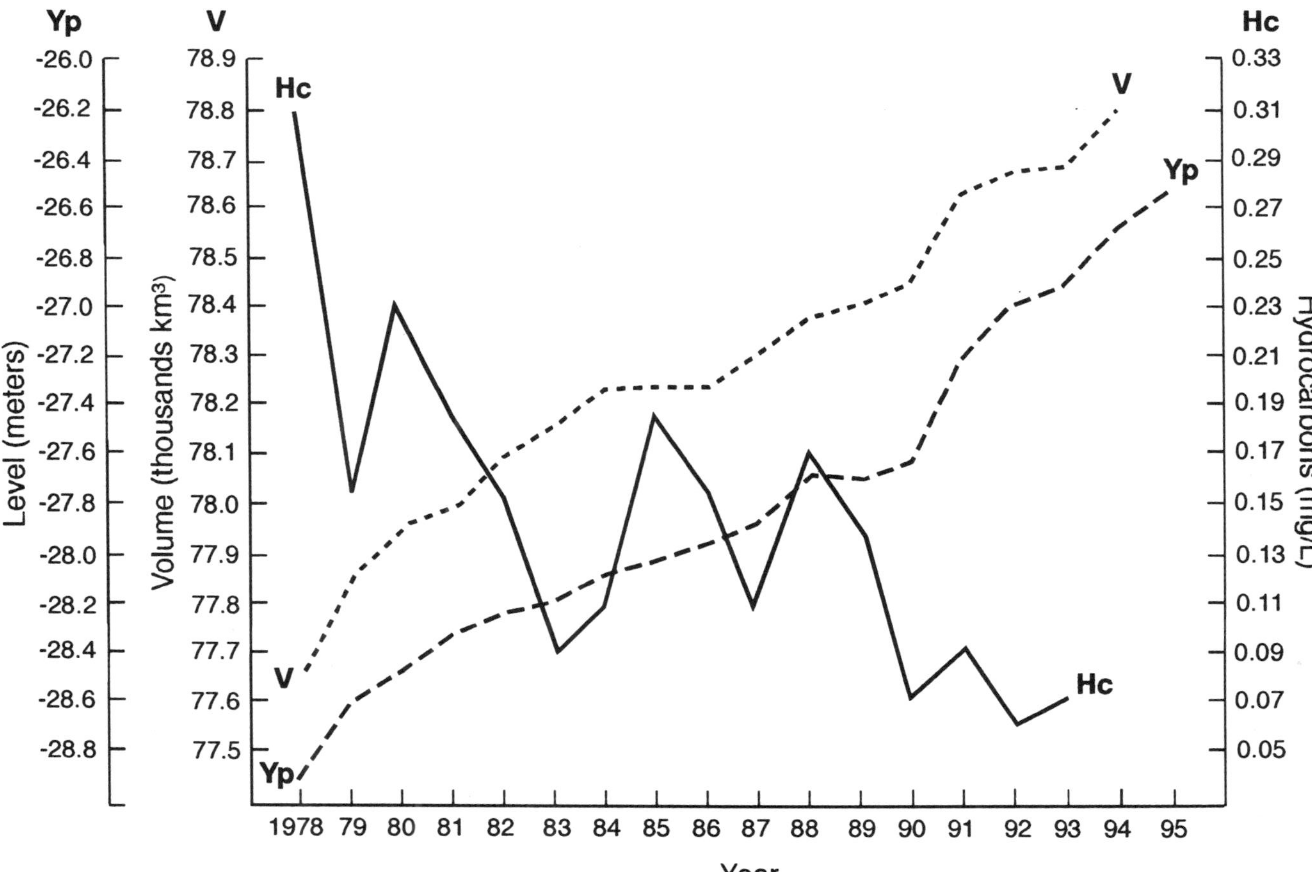

Fig. 1. Dynamics of level (Yp), volume (V), and concentration of hydrocarbond (Hc) of Caspian Sea (1978-1995).

TABLE 2. Changes in the water budget components of the Caspian Sea

Years	Sea level (m)	River flow (km^3)	Precipitation (cm)	Evaporation (cm)	Flow into Kara-Bogaz-Gol Bay (km^3)
1900-1929	−26.18	335.7	17.3	96.7	21.8
1930-1941	−26.80	268.6	18.5	100.4	12.4
1942-1969	−28.18	285.4	20.0	96.4	10.6
1970-1977	−28.64	240.5	24.3	103.9	7.1
1978-1985	−28.35	304.5	22.2	92.8	1.8
1900-1985	−27.35	298.2	19.4	97.4	13.6

Blocking Caspian Sea flow into the Kara-Bogaz-Gol Bay in 1980 reduced the outflow part of the water budget by 10 km^3 a year and has added to the increase in sea level of 2.5-3.0 cm annually. Between 1985 and 1992, an adjustable inflow of sea water (1.8-2.0 km^3 per year) into the bay was made. The actual outflow of water into the Kara-Bogaz-Gol Bay in the period between 1979-1990 has been about 75 km^3 instead of the forecasted 150 km^3.

As a result of river inflow and atmospheric precipitation between 1978 and 1985, the sea received almost 50 km^3 per year more water than was spent for evaporation and outflow to the Kara-Bogaz-Gol Bay. This led to an increase in sea level of about 13 cm per year and has caused its rapid rise.

Change in the level of the Caspian Sea has become a subject of research, mainly because of its impacts (often adverse) on the economies of the littoral states. Table 3 [8] shows how the change in level affects the surface area of the Caspian Sea.

TABLE 3. Change in sea surface area associated with level changes.

Level (m)	Surface Area (x1000 km^2)	Area Increase (x1000 km^2)	Surface Area Increase (%)
−26	401.0	9	2.3
−27	392.1	—	—
−28	376.5	15.6	4.2
−29	357.0	19.5	5.5
−30	340.1	11.9	3.4

The present sea level is −26.55 m and the increase in the sea surface area from 1979 (−29 m) has been about 40,000 km^2.

According to the data of Azerbaijan's Meteorology Committee, there has already been an inundation of several petroleum deposits, 600 km of coastline with a loss of 20,000 hectares of agricultural fields, 50 small cities and settlements, 250 industrial buildings, and railways and highways. With a continued increase in level up to −26.0 m, the flooded agricultural area will increase to 100,000 hectares, will inundate more than 18 settlements, a large factory manufacturing deep-water oil platforms, as well other buildings. Total damage at present is estimated to be more than US$2 billion.

Ecological Problems in The Caspian

Even if the problem of Caspian Sea level changes depends only slightly on human activity, another problem — its ecological condition — is completely determined by people. At present, pollution of the Caspian has reached catastrophic proportions. Large cities, industrial centers, seaports, oil terminals, transport, communications and so forth are located in the Caspian basin. All Caspian states have highly developed industries, agriculture, transportation and public services. As a result, the rivers running into the Caspian carry plenty of polluting substances. In addition the cities located on the Caspian coast dump their municipal waste directly into the sea.

One of the main polluting substances of the sea is oil. The technologies used in oil production in the marine environment are imperfect and dangerous. As a result, the Caspian Sea is polluted by petroleum, a loamy solution (used in drilling the wells), chemicals, as well as less pure waters used for industrial purposes. During oil exploration, extraction and transportation, accidental spills can occur [9]. Oil spills that occurred between 1978 and 1989 on Azerbaijan's offshore oil fields in the Caspian Sea are shown in Fig. 2. Nevertheless, their impact on the level of pollution of the sea is insignificant because of the dynamic hydrometeorological conditions and the relatively small spatial scale of the oil spills when compared to the total surface area of the Caspian Sea.

In recent years the pollution of the sea by organic substances has grown sharply. The polluting substances in the sea appreciably exceed permissible sanitary and ecological levels. The increasing amounts of pollution of the Caspian require comprehensive research with the objective to develop strict measures to prevent further pollution. Complex ecological systems can function normally within certain limits of variations of their parameters. A big departure from these limits can lead to the loss of stability of a system, if not its destruction.

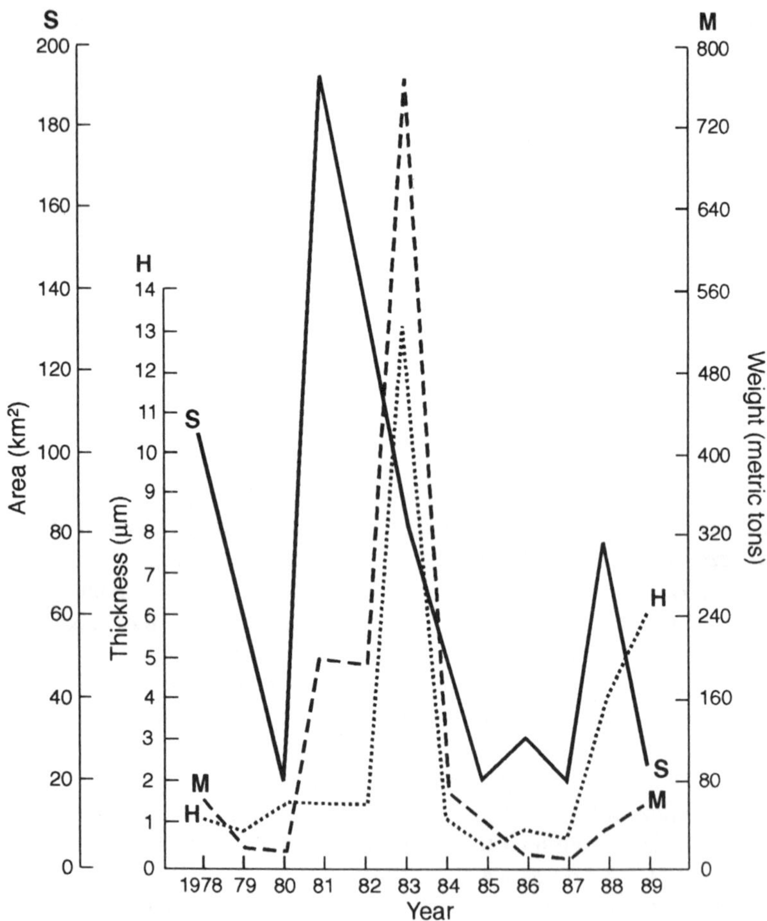

Fig. 2. Annual changes of size of oil spills in Azerbaijan "aquatorium" of the Caspian Sea. S=area (km^2); H=thickness (μm); M=weight (metric tons).

The main polluting substances of the surface waters of the Caspian are oil and oil products, phenols, synthetic surfactants, heavy metals, and pesticides (especially DDT). According to data for the years 1978-1992, the Caspian received an annual average of the following quantity of polluting substances (in thousands of metric tons): oil — more than 100; phenol — up to 1; synthetic surfactants — more than 3; copper and zinc — 9; as well as quantities of other heavy metals, pesticides, sulfuric acid and polluting substances. As a result, the quantities of hydrochemical pollutants in the sea [9] went beyond their allowable concentrations (Table 4).

TABLE 4. Hydrochemical parameters of Caspian Sea pollution (yearly averages for 1978-1991)

Components mg/l	Northern Caspian	Central Caspian	Southern Caspian	Average in the Sea
O_2	9.4	10.1	9.8	9.8
Hydrocarbons	0.16	0.10	0.13	0.12
Phenols	0.008	0.005	0.008	0.007
Synthetic Surfactants	0.05	0.05	0.06	0.05
NH_4	40.5	84.5	90.4	84.5
Pesticides	0.03	—	—	0.03
Copper	—	2.9	2.5	2.5
Mercury	—	0.04	0.05	0.05

Some of the sources of pollution are rivers [12], particularly those concentrated along the western coast of the Caspian Sea (see Table 5).

TABLE 5. Annual flow of various pollutants in the Caspian Sea 1981-1986

River	Annual flow (km^3/year)	Oil	Phenols	Synthetic Surfactants	Copper	Iron	Manganese	Mercury
Kura	14.4	735.9	82.0	466.6	52.7	39.1	44.4	1.44
Samur	1.6	141	—	8.27	9.6	9.8	8.9	—
Sulak	3.6	321.6	12.2	98.3	2.7	4.0	6.21	—
Terek	5.1	432.9	24.9	149.7	33.7	26.5	55.4	—
Total	24.7	1631.4	119.1	722.87	98.7	79.4	114.9	1.44

TABLE 6. Annual flow of pollutants into the Caspian Sea (1978-1992)

Pollutants	Thousands t/yr											
	Sea	%	with Volga flow	%	Azerbaijan	%	Turk-menistan	%	Dagestan	%	Kazakstan	%
Oil	105.60	100	89.3	84.6	15.870	15	0.320	0.3	0.065	0.06	0.01	0.01
Phenols	1.12	100	1.05	93.7	0.061	5.4	—	—	0.004	0.36	0.001	0.09
Synthetic Surfactants	6.20	100	4.1	66.1	2.000	32.3	0.012	0.19	0.01	0.16	0.08	1.3
Pesticides	1.92	100	1.58	82.3	0.334	17.4	—	—	0.003	0.03	—	—
Metals	9.20	100	8.1	88.0	0.820	8.91	0.020	0.02	0.03	0.33	0.23	2.5
Total	124.04	100	104.13	83.95	19.080	15.4	0.352	0.28	0.112	0.09	0.321	0.26

The Volga is a major source of pollution of the Caspian. Using data based on state statistical reports, the Volga along its length to the sea brings up to 23 km^3 of industrial and municipal flows containing 387 metric tons of weighted substances, significant quantities of phenols, synthetic surfactants, petroleum, heavy metals, pesticides, etc. Therefore, the delta of the Volga in the northern Caspian is the most contaminated.

The differences between pollutant flows from the Russian Federation and from Azerbaijan have become more substantial in recent years. This can be explained by the substantial decrease in industrial production in Azerbaijan.

Comparing Table 6 and Table 7 shows the amount of polluting substances going into the Caspian Sea from the Volga and from Azerbaijan's territorial waters.

TABLE 7. Inflow of pollutants in the Caspian Sea from the Volga and from the Azerbaijan part of the Sea (1990)

	All Sea	------Volga Flow-----		-------Azerbaijan Flow-------	
Pollutants	metric tons (x1000)	metric tons (x1000)	% of total	metric tons (x1000)	% of total
Oil	149.7	144.76	96.7	4.233	2.9
Phenols	1.4	1.232	88.0	0.192	3.7
Synthetic Surfactants	4.0	3.08	77	0.853	21.3
Pesticides	2.5	2.464	98.56	0.028	1.1
Metals	18.76	8.624	45.97	0.415	2.2
Suspended matter	17295.4	8008.0	46.0	33.375	0.2
Total	17630.36	8168.16	46.33	39.096	0.22

There has been a gradual decrease in the concentration of harmful components in the water and at the sea bottom. Nevertheless, despite the relative improvement in the quantitative indicators of changes in the environment (Figs. 2, 3, 4, 5, 6), the current ecological conditions of the region continue to be unfavorable.

Among the most polluted regions of the sea is the Apsheron Peninsula, and the different industrial and municipal sources such as the large centers of oil production of Baku and Sumgait. Pollution also comes from the northern regions of the sea, as a result of the prevailing southern current along the western coast of the middle stretch of the Caspian.

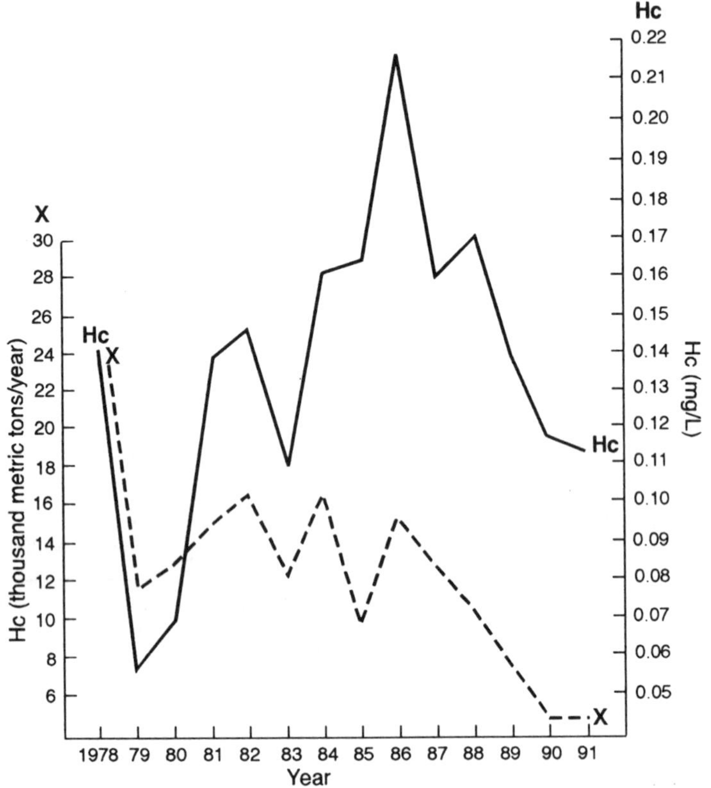

Fig. 3. Yearly changes of the concentration of hydrocarbons in Azerbaijan's "aquatorium" (territorial waters) of the Caspian. Annual averages, 1978-1991.

TABLE 8. Quantity of pollutants flowing into the Azerbaijan part of the Caspian Sea (annual data, 1977-1991), x 1000 metric tons

Pollutants	Republic	Baku	Sumgait
Oil	12.0	9	2.512
Phenols	0.063	0.025	0.018
Synthetic	2.75	0.340	1.66
Pesticides	0.334	—	—
Metals	0.648	0.221	0.427
Organic matter	148.470	99.078	24.167
Suspended matter	226.47	219.662	6.802
Acids	17.137	0.221	16.916

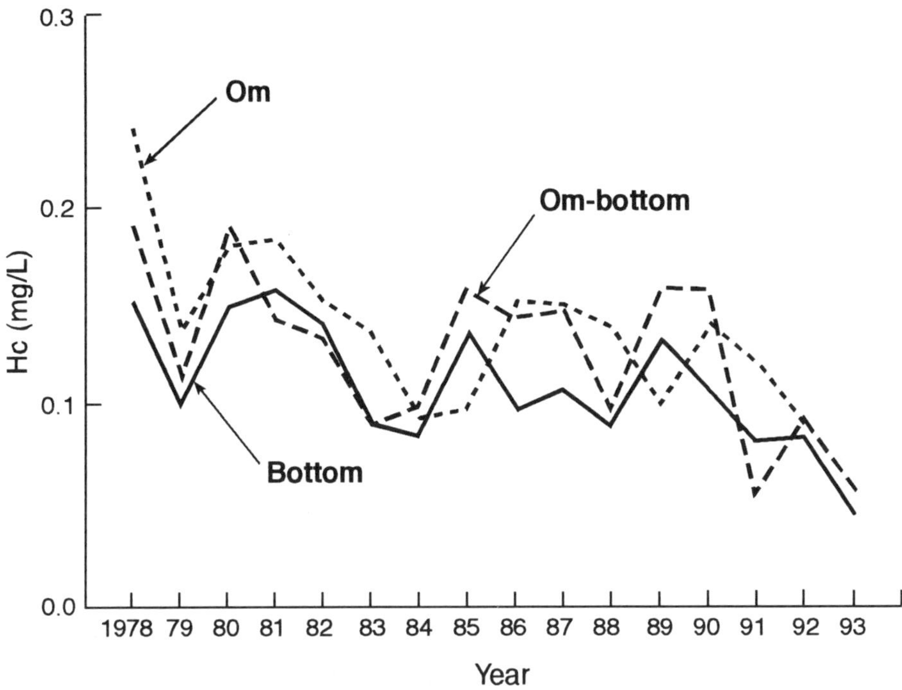

Fig. 4. Dynamics of hydrocarbon concentration (annual averages, mg/L, in Caspian Sea.

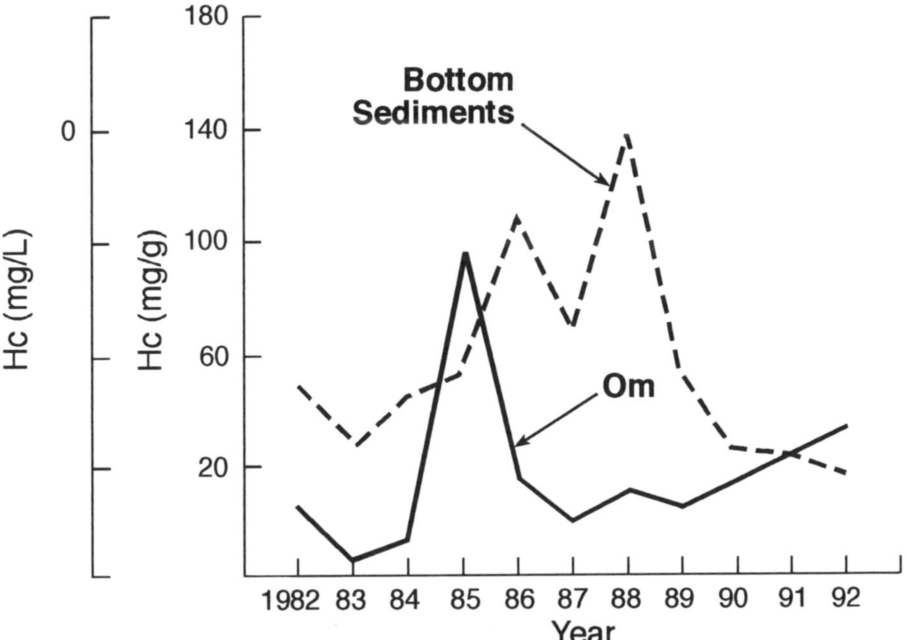

Fig. 5. Yearly changes of hydrocarbon concentrations in the bottom sediments of Baku Bay.

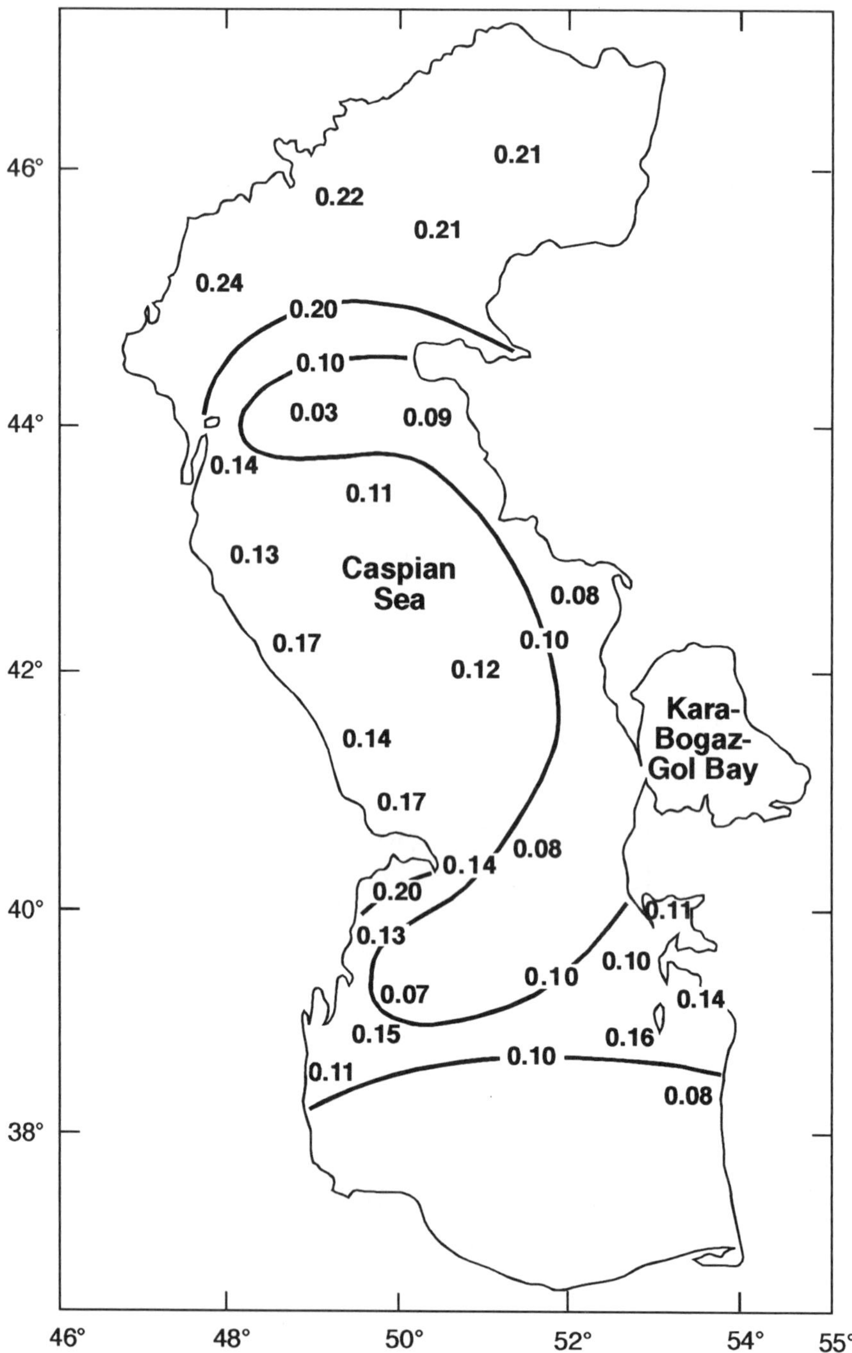

Fig. 6. Distribution of hydrocarbons (mg/L) in Caspian "aquatorium" (1978-1992).

The pollution of the Caspian has caused the outbreak of fish diseases and has spoiled the quality of those that survived. Heavy metals such as copper, zinc and lead accumulate in the flesh of sturgeon; pesticides accumulate in fish liver. Marine pike have almost completely disappeared [10], and the quantity of Caspian herring has sharply decreased. Figure 7 shows the correlation between oil pollution and the amount of sturgeon landings [13].

Since the early 1990s, most of the regular observations of the pollution in the territorial waters of all the littoral countries were stopped. Thus, unfortunately, at present the opportunity for an evaluation of the ecological situation in the Caspian is complicated.

Recent Assessments by Azerbaijan's Space Agency

The following works have been undertaken recently by the Azerbaijan National Aerospace Agency.

1. *Study of the variability of hydrometeorological conditions of the Caspian Sea.*
   In this work, an investigation was conducted of seasonal and synoptic variability of temperature fields of a surface layer of water, the near-water layer of atmosphere, wind-driven waves, humidity and other hydrometeorolgical parameters.

2. *Study of the processes of turbulent transfer in a boundary "air-sea" layer.*
   An investigation was conducted of the physical processes of small-scale air-sea interactions that influence remote sensing measurements. Measurements of the characteristics of the "air-sea" boundary system were carried out on an experimental basis from a stationary platform in the sea (at a depth of 40 m).

3. *Study of hydrophysical, hydrochemical and hydrobiological fields of the Caspian Sea for the purpose of creating remote sensing test sites.*
   Complex research was carried out on the physical, chemical and biological aspects of the Caspian, using data both of hydrometeorological observations and of synchronous satellite experiments with application of in situ and remote sensing techniques. Research ships and planes were used in these experiments.

4. *Development of an atlas of the Caspian Sea.*
   Based on data of hydrometeorological observations for 1900-1990, a detailed atlas of the Caspian Sea was developed. The atlas was also

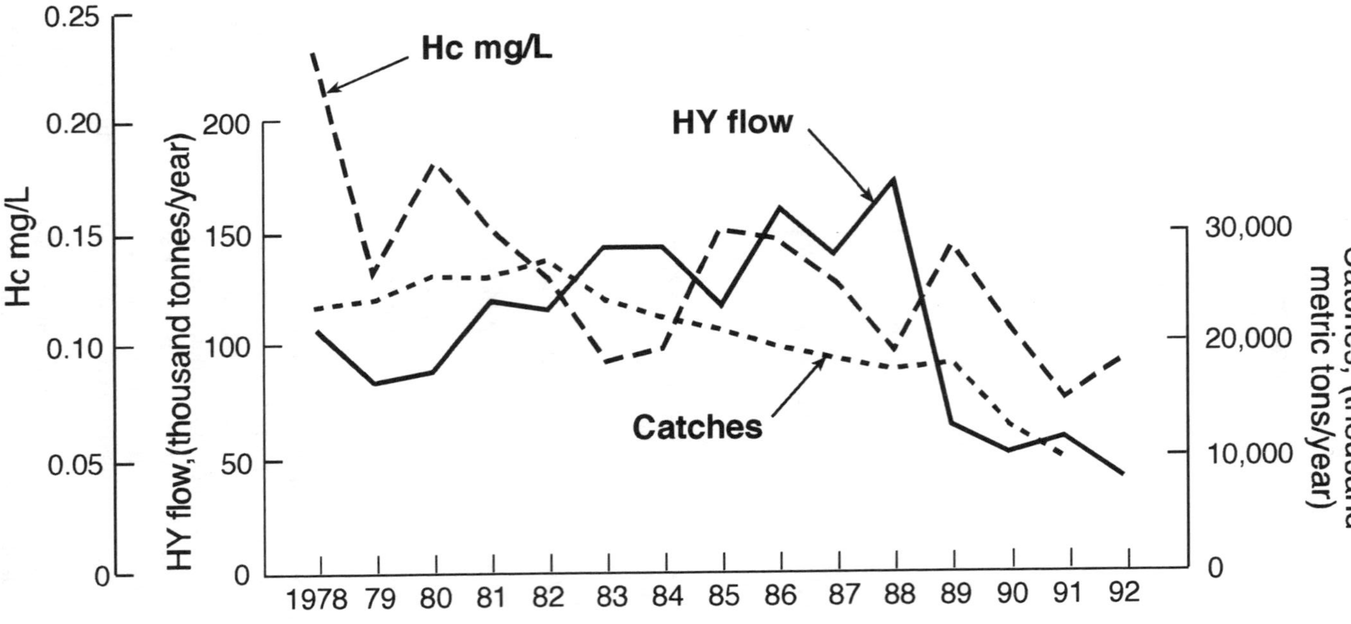

Fig. 7. Yearly dynamics of flow (thousand metric tons), concentration (mg/L) of hydrocarbons (bottom), and sturgeon fish catches (thousand metric tons) in Caspian Sea (1978-1992).

based on the materials from various expeditions conducted by specialists from the Azerbaijan National Aerospace Agency and other organizations. The atlas contains data on both hydrometeorological and ecological parameters. Monthly average maps of the main hydrometeorological parameters (e.g., wind speed, temperature of water and air, humidity, etc.) were constructed as were maps of the distribution of the most dangerous polluting substances (e.g., oil, phenol, heavy metals, synthetic surfactants, etc.).

5. *Study is now being undertaken of the ecological condition of a "land-sea" boundary in connection with fluctuations of the level of the Caspian Sea, using materials received by six-channel photo-camera MKF-6 from aboard an AN-30 plane and ground expedition activities.*

Monitoring Needs for the Caspian

The Caspian Sea is a unique object of nature and the necessity for its protection requires the acceptance of joint measures by countries in the region. It is necessary to restore a central and uniform monitoring of the Caspian environment and to develop a comprehensive program of ecological monitoring of the Caspian Sea. Such a program should include:

1. *Study of a time-and-space structure of the physical, chemical and biological aspects as well as aspects of polluting substances.*
   For this purpose a major use of archival and published materials and the results of expeditions and experiments are required. Such information is necessary for the creation of a data bank, which will enable the collection and distribution to all littoral countries of marine parameters under all weather conditions.

2. *Development of diagnostic and predictive models of the sea and atmosphere above it.*
   These models are necessary for the analysis and forecast of physical processes in the sea, including processes of air-sea interaction. They are especially important for the development of measures needed to decrease pollution.

3. *Creation of effective technical means of in situ and remote-sensing measurements.*
   For environmental monitoring it is necessary to organize test sites, enabling the calibration of remote-sensing techniques.

In situ measurements have much accuracy, but they cannot be extrapolated effectively to a large area; remote measurements can monitor large areas. In a general system one should include local, regional, national and international subsystems. Each subsystem determines problems of appropriate scales.

In sum, monitoring changes in the Caspian environment should be one of the highest priorities of countries that share the Caspian coastline and sea. Recently, Kosarev and Yablonskaya [14] stated the need in the following way:

> For evaluation of the state of the Caspian ecosystem and the timely interpretation of the changes occurring it is essential to keep the sea regime under constant observation by introducing a complex system of monitoring of the natural milieu, involving wide-scale stationary and expeditionary observations. Only in this way will it be possible to obtain the data necessary to resolve the most important problems of the Caspian Sea, and to work out long-term forecasts and the complex of measures necessary to protect the water body and to promote a balanced use of its natural resources.

Concluding Comments

The Caspian is a unique natural ecosystem and study of it by experts from each of the littoral states is required. Unfortunately, at present, the economic conditions of these states do not provide hope for systematic supervision of the sea or for the widespread acceptance of measures to improve its ecological condition. Responses to changes in the Caspian Sea require the participation and assistance of international organizations. Attempts at such international cooperation have been made. For example, in Tehran in 1994 the Coordination Committee on Hydrometeorology and Monitoring of Pollution of the Caspian Sea was created and a comprehensive regional program was developed under the aegis of the World Meteorological Organization. At a meeting of experts from Caspian countries a detailed plan of action was developed, with the financial help of international organizations.

In 1995, experts of littoral states participated in comprehensive hydrologic, hydrochemical and hydroradiation research on the Caspian, along with the International Agency on Atomic Energy, Paris and Strasbourg Universities, and the Marine Laboratory of Monaco. The research was conducted during one month, using a research vessel of the Azerbaijan Committee on Hydrometeorology.

These are the first positive steps, following several years of a break in collaboration among the littoral states. It rekindles hope for achieving regional observations and a study of the Caspian Sea which could favorably influence the life of many millions of people in the region.

References

1. Voropayev, G. (Ed.), 1986: *The Caspian Sea: Hydrology and Hydrochemistry*. Moscow, Russia: Nauka, 262 pp.

2. Aliyev, Sh.D., 1954: A level of the Caspian Sea on historical maps. In: *Fluctuations of the Level of the Caspian Sea*, Vol. 2. Moscow, Russia: Academy of Sciences of the USSR, 110-117.

3. Berg, L.S., 1960: Level of the Caspian Sea in historical times. In: *Selected Proceedings*, Vol. 3. Moscow, Russia: Academy of Sciences of the USSR, 281-326.

4. Fedorov, P.V., 1956: Reasons for the fluctuation of the Caspian Sea level in a quarter period. In: *Fluctuations of Caspian Sea Level*. Vol. XV. Moscow, Russia: Transactions of the Institute of Oceanology, Academy of Sciences of the USSR.

5. Rychagov, G.I., 1977: *Pleistocene History of the Caspian Sea*. Doctoral thesis. Moscow, Russia: Moscow State University.

6. Zektser, M.S., R.G. Jamalov, and A.V. Meshetejan, 1984: *Underground Water Exchange between Land and Sea*. Leningrad, Russia: Nauka.

7. Kosarev, A.N., and F.B. Makarova, 1988: About the changes of the level of the Caspian Sea and the possibility of forecasting it. *Proceedings, Geographical Series,* **5**. Moscow, Russia: Moscow State University.

8. Kritski, S.N., D.V. Korenistov, and D. Ja. Ratkovich, 1975: *Fluctuations of Caspian Sea Level*. Moscow, Russia: Nauka, p. 157.

9. The Department of Hydrometeorology and Control of Environment of Azerbaijan, 1991: *The Yearbook of Water Quality of the Caspian Sea, 1978-1991*. Baku, Azerbaijan: Azerbaijan Department of Hydrometrology and Control of Environment.

10. Gul, A.K. et al., 1994: Dynamics of pollution in Azerbaijan aquatorium of Caspian: Current ecological problems and methods of solution. *First International Scientific and Technical Conference*. Baku, Azerbaijan: Azerbaijan National Aerospace Agency, p. 93.

11. Afanasjeva, N.A., and S.N. Kiryanov, 1991: A condition of pollution of waters of the Caspian Sea at the present stage. Materials of an All-Union Meeting on Problems of the Caspian Sea. Gur'ev, Kazakstan, p. 31.

12. Gul, A.K. et al., 1989: The problem of surface and underground waters in the Caspian region. *Problems of Socioeconomical Development*, Vol. 5, Water and Industrial Problems. Moscow, Russia: VINITI, p. 83.

13. Kasimov, A.G., 1994: *Ecology of the Caspian Lake*. Baku, Azerbaijan: Azerbaijan Publishing House.

14. Kosarev, A.N. and E.A. Yablonskaya, 1994: *The Caspian Sea*. The Hague, The Netherlands: SPB Academic Publishing, p. 259.

ECOLOGICAL, SOCIAL AND ECONOMIC PROBLEMS OF THE TURKMEN CASPIAN ZONE

AGADJAN G. BABAEV
Desert Research Institute
Ashgabat, Turkmenistan

Turkmenistan is one of the five independent states bordering the Caspian Sea. The Caspian Sea basin is a unique natural region with great significance for the economy of the country and for interstate relations. The sea has great influence on regional climate. It smooths the amplitude of fluctuations of the temperature of the whole Caspian region air mass, thereby considerably "softening" the climate. Supplying the atmosphere with an enormous amount of moisture from evaporation, the sea mitigates the influence of the hot deserts of Central Asia. However, the recent rise of the Caspian Sea level, which started in 1978, is fraught with negative ecological, social and economic consequences.

The total length of the Turkmen part of the Caspian coastline from north to south is about 1200 km and extends from its northern border with Kazakstan to its southern frontier with the Islamic Republic of Iran. The Caspian coastline is considerably jagged and there are the Kara-Bogaz-Gol and Krasnovodsk Bays and the Krasnovodsk and Cheleken Peninsulas (Fig. 1).

As the administrative unit, the Turkmen coastal zone is included in Balkan Velavat (the former Krasnovodsk Oblast) which occupies 28% of Turkmenistan's territory. Its population is nearly 500,000. The labor force of Balkan Velavat is being developed for the petroleum-gas and chemical industries. Therefore, a significant part of the population is engaged in the natural resources and industrial sectors (e.g., extraction and processing of oil, and of chemical raw materials, fishing, and the like). Such industrial centers as Turkmenbashi, Nebitdag, Okarem, Koturdepe, Bekdash, Cheleken and other towns have developed on the basis of the exploitation of mineral and other primary resources. Agriculture in the region, because of the lack of fresh water, is characterized chiefly by animal husbandry.

The history of cyclical fluctuations of Caspian Sea level has been known for a long time. The causes of this phenomenon in the past as well as in the present,

M.H. Glantz and I.S. Zonn (eds.),
Scientific, Environmental, and Political Issues in the Circum-Caspian Region, 97-103.
© 1997 *Kluwer Academic Publishers. Printed in the Netherlands.*

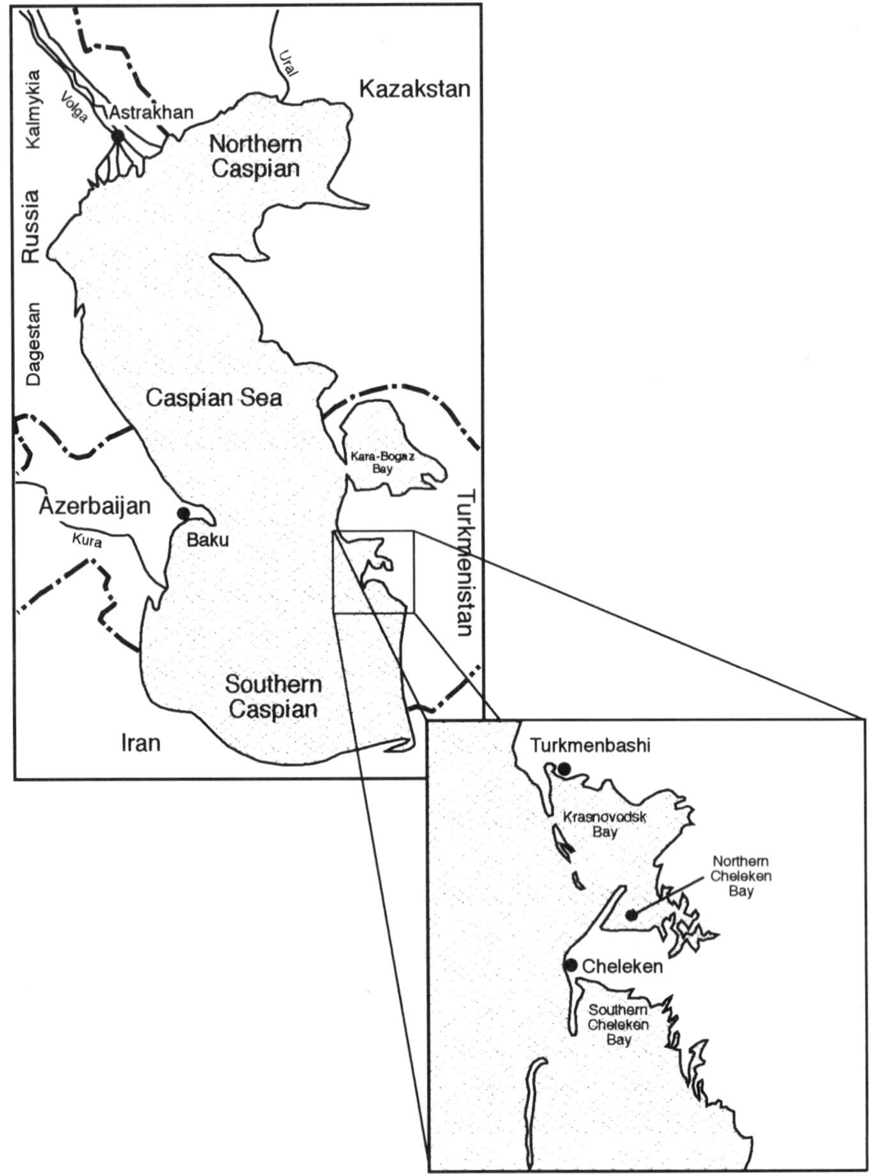

Fig. 1. Turkmenian coast of the Caspian Sea

however, do not appear to be obvious. There are two different, contending hypotheses to explain the cause of repeated fluctuations of sea level. Supporters of the climatic hypothesis explain the increase in Caspian Sea level as the result of a planet-wide rise in atmospheric temperature which is accompanied by an increase of sea and ocean evaporation leading, in turn, to an increase in atmospheric precipitation on continents as well as within the Caspian Sea basin. The tectonic hypothesis pertains to the sea level's behavior that is affected by periodic geological processes of internal strain shifts (e.g., compressing and stretching) of the earth's crust leading either to non-tectonic shifts of the seabed and shore, or to the considerable fluctuations in the amount of underground drainage. Neither hypothesis, in our opinion, allows us to make reliable forecasts of Caspian Sea level behavior. In the first case the difficulty in evaluating sea level behavior is stimulated by the lack of sufficient well-grounded methods of forecasting fluctuations or changes in regional climate on the one hand, and by the inaccurate measurements of the Caspian water balance components on the other.

Several researchers have recently attributed the increase in sea level to the 1980 damming of the strait that connects the Caspian Sea with the Kara-Bogaz-Gol Bay. In June 1992 this dam was dismantled and Caspian Sea water was once again allowed to flow freely into the bay. In the first months the drainage volume totaled 700-720 m^3/sec or 1.7 km^3/month. In 1993, due to the erosion of some coastal areas of the bay, the drainage volume increased to 800 m^3/sec or 2.0-2.2 km^3/month, and in 1994-1995 it increased to 1200-1500 m^3/sec or 3.7-4.1 km^3/month. During 3.5 years, about 85-90 km^3 of Caspian Sea water entered the bay, undoubtedly lowering the rate of sea level rise. Nevertheless, as measured at Bekdash and Kara Bogaz stations, the sea level rose within the 1992-1994 period by 60 cm from a level of -27.22 m (NB: These numbers are taken only from the minimal average monthly figures of level at the beginning and at the end of each year). If seasonal fluctuations were to be taken into consideration, the average annual increase would be greater. In other words the actual Caspian sea level increase, including the free drainage of its waters into Kara-Bogaz-Gol Bay, exceeded the forecasted increase by nearly 5 times. The dynamics of filling up the bay is shown in Fig. 2.

With regard to the tectonic hypothesis it should be mentioned that in spite of the rather reliable fixed oscillatory movements using seismic data, the scale of induced changes of the Caspian Sea volume resulting from the amount of underground drainage in the discharging zone can be evaluated only with the use of indirect (or proxy) data.

We consider the Caspian region to be the integral natural "geosystem" where endogenous, exogenous and solar factors closely interact. Their interactions are the real source of many natural changes, disasters and catastrophes, including sea-level fluctuations.

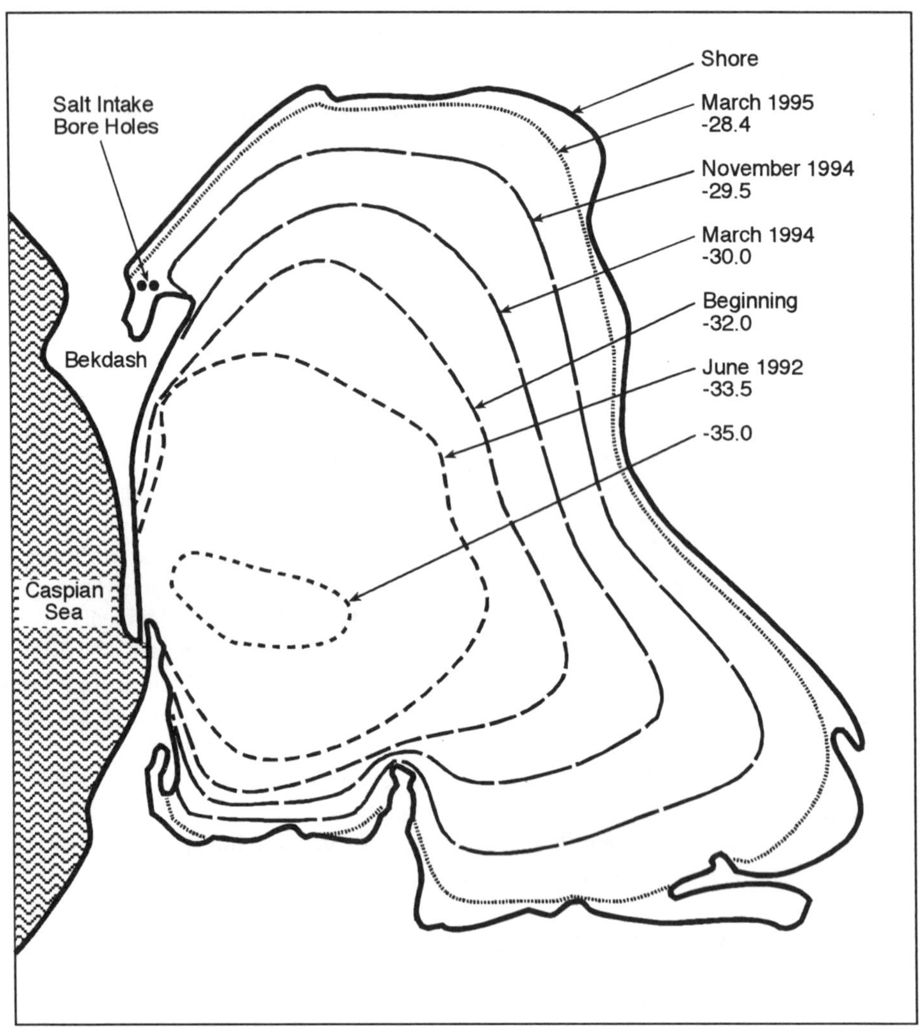

Fig. 2. Dynamics of the filling up of the Kara-Bogaz-Gol Bay

The Caspian Sea level increase of the past 18 years has already inflicted considerable losses in the economic and social sectors in the region. For example, within the South-Cheleken spit, the creation of Dervish Island is in process; sea water movement is taking place both from the open sea and from the South Cheleken Bay. Buildings and infrastructure of the lower part of the Cheleken town are being flooded as well as the Karagel settlement and other important sites. Water also inundated the oil-industry workers' holiday zone on Cape Khelles. All communications are submerged: electrical transmission links, oil and gas mains, water mains on the Cheleken-Koturdepe-Djebel route. Large oil deposits of Western

Cheleken, Komsomolsk (Oval-Toval) and Koturdepe are exposed to intensive inundation.

Sea-water pollution by oil products has resulted from the poisonous drift of waste from derricks, oil barns and burial grounds. The rapid process of separation of Cheleken from the continent is taking place along the Esheklyar saddle (The Skikh-Dervish natural boundary (Fig. 3)).

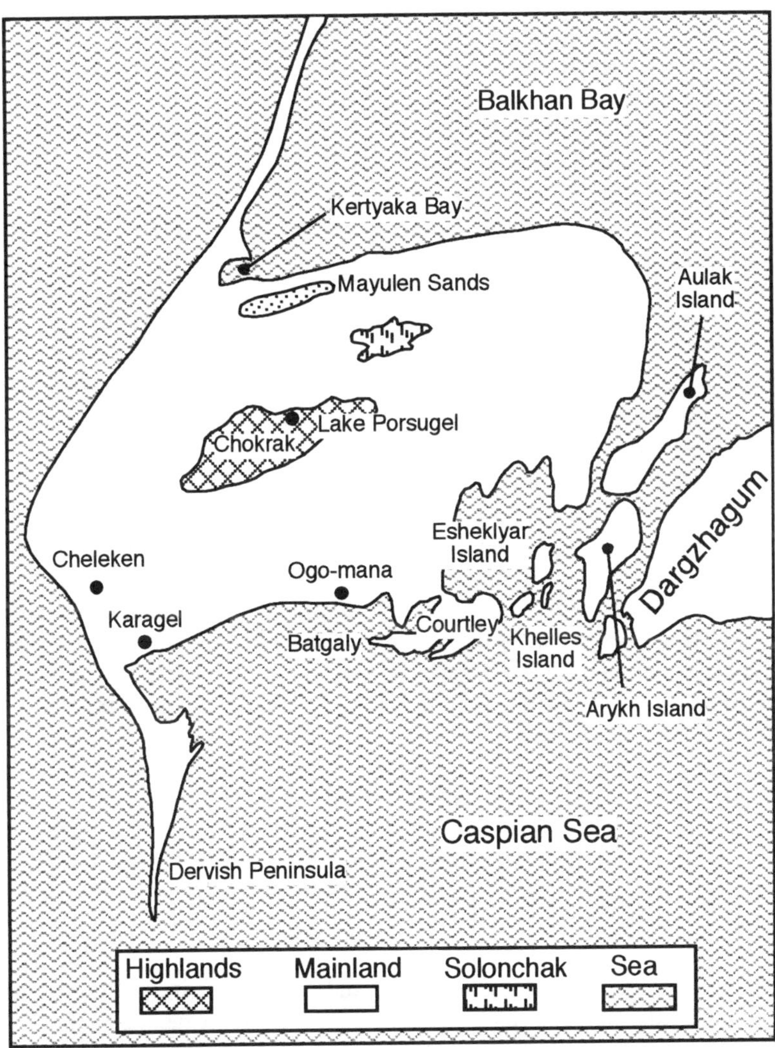

Fig. 3. Cheleken Island and Dardzhagum in 1909 (after Veber and Kalitskiy)

The first real features of the quick transformation of the Cheleken Peninsula into an island were the destruction in several places of the Cheleken-Koturdepe highway by wind-driven wave action in June 1995 and, for the first time in the last 50 years, the joining of sea water of the North and South Cheleken bays.

In the history of the Caspian Sea, the present increase in level is a manifestation of variability which will last, according to our preliminary prognosis, until 2015-2020. As a result, sea level will rise by another 2-2.5 meters. If that were to occur, that would have the following serious negative social, economic and ecological consequences:

- The greater part of oil and gas mains will be submerged, and will be put out of commission because of corrosion caused by sea water;
- The large industrial enterprises in Turkmenbashi, Cheleken, Bekdash, Djantha and other towns, where the greater part of enterprises of oil, oil-chemical, chemical industry and sites of the construction industry are concentrated, will be threatened;
- Housing in the coastal settlements will suffer considerably, as well as social and cultural and sanitation sites, adversely affecting life in the area;
- Great damage will occur to a number of oil and gas and other mineral resources deposits which will require enormous additional means for their protection;
- Fish-breeding farms will be destroyed.

It is quite obvious that the expansion of the area inundated by the Caspian Sea will provoke some changes in the climate along the coast. The changing landscape components, such as water and air masses, may lead to a different local climate. It is well known that in the Turkmen Caspian zone there are splendid beaches, mineral springs and medicinal volcanic mud. The warm sandy coast has been regarded as a potential recreational resource of international significance, as the period of an active bathing season is longer than that of the Black Sea and Azov Sea coasts.

The unique international Krasnovodsk biosphere reserve is located in the Turkmen Caspian Zone. It was founded in 1932 for the protection of waterfowl and birds living near the water: nearly 70% of all birds wintering at the Caspian Sea gather there.

In sum, the submersion of industrial and communal sites, buildings and mineral resource deposits will delay economic growth, and generate social difficulties. Incomes and living standards will decrease, and the reduction of work places will generate unemployment. Additional means will be required to create new industries and jobs.

Psychological factors are also important. A considerable part of the population will have to change professions as well as lifestyles. Migration of population from the submerged territories will increase as people search for new places to live. As previous experience shows, the most qualified laborers tend to migrate; this will require additional expenses for the training of new workers.

Further increases in Caspian Sea level may force structural reforms in other branches of the economy. In particular, it will lead to the development of the Cheleken-Okarem free enterprise zone. Nowadays, the government is planning to solve a number of social and economic problems by means of fostering the development of this zone.

Complex measures are directed at the protection of the most important sites of the sea coast and at the prevention of polluting substances and harmful waste from drifting to the sea or from being deposited directly into the sea.

To get full and objective information on the reasons for large level fluctuations and the forecast of their behavior, it is necessary to organize aircraft, satellite, geological and geomorphological monitoring of the coastline and the ecological situation by way of a unified interstate program on the Caspian Sea.

Conclusions

1. Revealing and substantiating the reasons for long-term fluctuations of the Caspian sea level is considered to be a rather difficult scientific problem. Existing competing hypotheses, e.g., climatic and tectonic ones, do not provide the basis for a reliable forecast of sea level behavior. Meanwhile, information on the problem shows that, from either a climatic or a tectonic point of view, a decline in Caspian sea level is not to be expected in the near term. The evidence at hand suggests a possible level increase of 2-2.5 meters or more.

2. Unfortunately, the states in the Caspian region do not yet have a unified action program for sea level stabilization. Isolated, disparate attempts to protect coastal sites by different measures, such as the reinforcement of vulnerable shorelines, may only delay their submersion for a short period of time.

3. A more attractive measure at the regional scale is to let Caspian Sea water flow into the "dry" depressions along the Eastern sea coast at Kara-Bogaz-Gol Bay (Turkmenistan), Ash-Sor, Karagie depressions and the "Komsomoletz," Sor-Kavdak Bays (Kazakstan).

THE PROBLEM OF THE CASPIAN SEA LEVEL FORECAST AND ITS CONTROL FOR THE PURPOSE OF MANAGEMENT OPTIMIZATION

GRIGORIY V. VOROPAYEV
Research Center "Kaspiy"
Scientific Coordination Center
Moscow, Russia

Natural Resources of the Sea and Its Coastal Zone

The Caspian Sea region is exceptionally rich in natural resources — biological, agroclimatic, balneological, and recreational, and the aggregate of such resources makes this region favorable for human settlements and for agricultural and industrial production. Under conditions of a stable political situation, the economies of the region could, in the near future, enhance their development prospects. Oil and gas deposits are the region's main mineral resources. There are nearly thirty deposits now explored in the coastal zone of all the littoral states and directly in the sea, in addition to those already developed. In any event, the national wealth of the states in the Caspian region will likely increase rapidly with the exploitation of its known natural resources.

Extraction of oil in this region started near the city of Baku (Azerbaijan) as early as the eighteenth century. Small-scale development of the sea's deposits occurred at the beginning of the nineteenth century, and the first large industrial oil deposit was discovered in the sea in 1949. Since then, the Caspian Sea had for many years become one of the main oil suppliers in the USSR, providing about 10 million metric tons per year.

The Caspian Sea region occupies one of the world's major oil reserves, with reserves estimated at more than 18–20 billion metric tons. Inadequate study of the oil and gas reserves of the region suggests that projections could run significantly higher than original estimates. Oil extraction is currently carried out in small quantities, less than 20 million metric tons per year, half of which comes directly from the sea's oil deposits.

Other mineral resources in the region are stones for construction, including valuable types of trim stones and various salts, and salt deposits in and around the Kara-Bogaz-Gol Bay.

Biological resources of the Caspian Sea have in the past been notably large and practically priceless. The largest stock of sturgeon in the world are located here. The

M.H. Glantz and I.S. Zonn (eds.),
Scientific, Environmental, and Political Issues in the Circum-Caspian Region, 105-117.
© 1997 *Kluwer Academic Publishers. Printed in the Netherlands.*

Caspian sturgeon catch makes up more than 90% of all catches of this species on the planet. In this respect the Caspian is unique and of importance for the humanity of the Earth.

The Caspian — its river systems, estuaries, and shorelines — is also rich in other species of fish, water fowl, and animals. After separation of this water body from Pont-Azov, remnants of fauna of the tertiary seas of the Sarmatian and Pontian periods were preserved and its range of species diverse. There are seals in the Caspian Sea, the only sea mammal. Large areas of fertile lands are concentrated in the sea's coastal zone, making it possible to cultivate agricultural crops. Of particular value are flood lands of the largest river basin, the Volga, and their estuarine area. Irrigated agriculture developed in ancient times in these regions, promoting the progress of other branches of economy and cooperative relations in the sea and in the coastal zone. Population density is high in the mouths of the rivers, where the largest cities are located. Several seaside areas are excellent places for recreation, tourism, and nursing homes.

Conditions for the Optimum Use of Natural Resources

The sea and its coastal zone represents a single natural complex, with diverse and very complicated transboundary interrelations that determine the reproduction and conservation of resources, as well as their quality and productivity. This must be taken into account by providing an objective assessment of the resources, as well as the conditions for the sustainability (i.e., preservation) of their abundance, reproduction, and quality.

There are now five states using the resources in the coastal zone, and their national self-interests are not necessarily complementary. Therefore, there is a need to establish a regional political mechanism guaranteeing conflict-free use of Caspian resources. The primary need in international resource exploitation is the maintenance of a stable political situation among the littoral countries *and* cooperation by the leaders of these states.

Theoretically, all natural resources could be considered as renewable, from the point of view of their formation and depletion in the process of evolution of the Earth, as displayed through geological, climatic, and biological processes. However, with regard to the time factor and in comparison with the period of human life (e.g., generations of twenty years or so), some of these resources become exhausted in a certain location (e.g., oil, gas, certain types of salts, living marine resources), while others can be viewed as renewable and sustainable (e.g., soil fertility, aquatic and terrestrial biological resources). In other words, there should be a better distinction between renewable and nonrenewable resources. Thus, our understanding of the optimal use of natural resources includes, first of all, preservation of conditions for the continuous reproduction of renewable resources and the coordinated, measured use of nonrenewable resources.

From an economic viewpoint, a demand arises for optimized costs and benefits. However, it is impossible to achieve this without specific planning and without an ability to forecast changes (e.g., climatic) in natural processes. More specifically, the most important factor in the use of natural resources in the Caspian region is the need for forecasting sea level for the specific periods of various economic activities (e.g., seasons, years, service life of erected structures).

A stability of the sea level is of great importance. Under stable conditions, we would know the sea level over the long term (with long-term variations less than 0.3–0.5 m, and small interannual changes of less than 0.2 m). Such an "ideal" variation in sea level within the last two centuries has occurred for periods of 10 to 15 years. However, during a major part of the last two centuries, the sea level regime has not met those ideal conditions, and has caused ecological changes and economic losses.

A mean annual sea level in the range of –26.5 to –27.5 m is the most favorable for the reproduction of biological resources in general and sturgeon stocks in particular. Previously calculated conditions had specified those extreme physical boundaries of fluctuations for the long term as –26.0 to –28.0 m. The decline in sea level below the –29.0 m mark has been identified by biologists as critical for the commercial stocks of sturgeon.

A lower sea level of about –27.5 to –28.5 m, which existed during the period of construction and modernization of major port facilities, railways and vehicle terminals in the coastal zone, is most suitable for navigation purposes. This level is also beneficial for the protection of the largest percentage of urban buildings, embankments, and recreation areas in the region.

The lower levels, which occurred before 1977 (i.e., up to –28.5 m), also enabled the use of land resources and estuaries for agriculture in the newly exposed part of the coastal zone. However, a further decline in the level, although increasing the exposed area, yielded no additional fertile land for cultivation or for pasture. To a greater extent, low sea levels meet the needs for the development of oil and gas mining activities in the sea and on the coast, by extending the opportunities for drilling operations on the land and islands, and making the exploitation of constructed platforms and other structures in the sea more reliable. Low levels also encourage the construction of urban infrastructure on the newly exposed seabed.

The mining of mineral and salt resources in the Kara-Bogaz-Gol Bay ended in 1954, and the use of natural brine from aquifers near the bay had also ended because they were no longer connected to the Caspian Sea. However, the sustainable production of salts in the bay, depending on the level of demand, is possible and technically trouble-free, under any sea level which occurred in the twentieth century.

Unfortunately, until the present, the needed assessment of experiences in the use of natural resources has not been carried out, making it impossible with present knowledge either to assess quantitatively the influence of the sea level regime on the ecology and economy of the region, or to identify the optimum level of exploitation of land and sea resources. Ecologically, a stable-level regime is important. Long-term or sharp increases in the level can adversely change the quality of the aquatic environment in the coastal zone, which is the most important link in the reproduction of biological resources. A similar continuous decline in the level causes changes in the water regime of coastal ecosystems and the reduction of biological productivity of soils and perennial plants. In this respect, studies that can identify appropriate measures for the efficient development of the coastal zone under conditions of a fluctuating sea level are of principal importance.

The History of Sea Level Dynamics

The sea level regime and changes in its mean annual position within the periods of one year to several years to decades, exerts a significant influence on the efficient use of natural resources of the sea and its coastal zone. A considerable amount of scientific research has already been completed on Caspian Sea level changes on geologic time scales. When the sea lost its connection to the Pont-Azov system, it was transformed into an inland water body, and its water budget became completely dependent on the inputs of regional rivers, precipitation, and evaporation. Available data point to the fact that, during the late Pleistocene and the Holocene (about 15,000 years ago), the sea level could have varied in level from –10 to –50 m. However, there is no information about the interannual dynamics of the level during that period. Taken as a whole, this information about sea level is of low reliability. Information about sea level is to some extent better for the last four centuries. During that period, the sea level fluctuated within a smaller range of –22 to –29.0 m. However, even for much of that period up to the middle of the nineteenth century, there is no direct information on the interannual changes in Caspian Sea level.

Adequate information about interannual changes in sea level is available only for the last 150 years, when regular measurements of the level were carried out, using several measuring devices. Different errors in measurements occurred during that period because of the replacement of the measurement systems, the movement of gauge sites, etc. The observation data were recently subjected to careful analysis, and a continuous reliable data series was obtained of the mean annual level and elements of the sea's water budget. Using these data along with other studies about the basin, quantitative assessments were made of the influence of economic activity on changes in the water budget and of the dynamics of sea level. A time series of reconstructed sea level was developed, i.e., for conditions as if there was no economic activity in the basin and the regime of the sea had been completed determined by climatic factors for the period under investigation. Figure 1 depicts mean annual values of observed and reconstructed sea level for the period of instrumental observations.

Without considering the features of annual sea level dynamics, the available observational data disclose the mechanisms of hydrophysical processes: the sea level rises with the increase of river-water inflow during floods; it declines because of increased summertime evaporation from the sea surface; river flow influences the dynamics of water

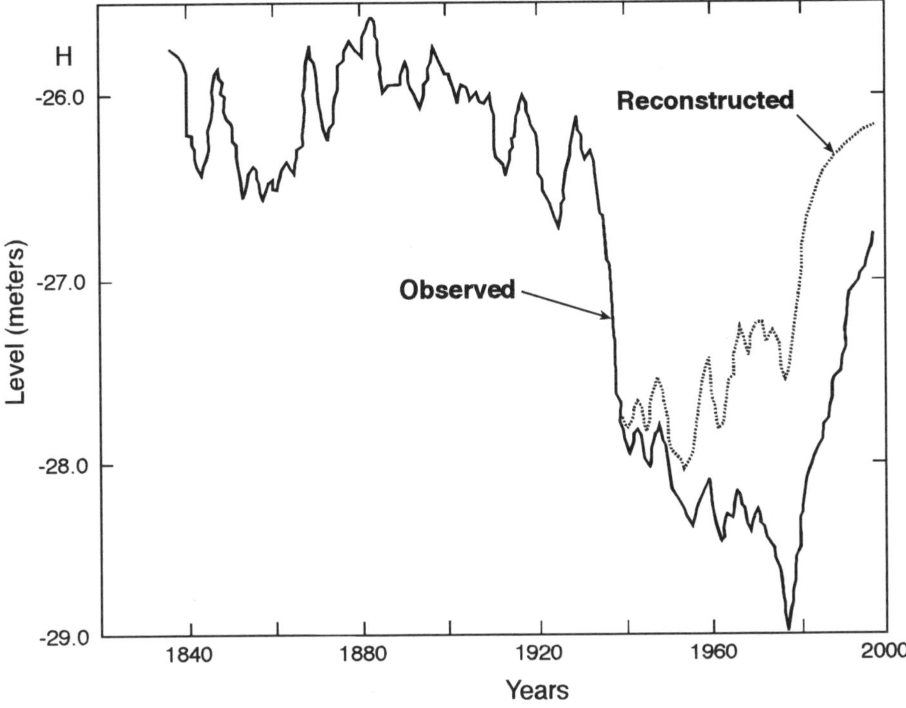

Fig. 1. Mean annual values of observed and reconstructed sea level

currents, mineralization processes, etc. The dynamics of annual and long-term sea level changes are governed by positive and negative annual increments to the water balance.

The amplitude of sea level fluctuations for the period of instrumental observations (1830 to present) was 3.8 m, with the highest recording in 1882 at –26.2 m and the lowest in 1977 at –29.0 m. In 1995, the mean annual level was –26.51 m. Conditionally, three time periods, having sharply distinct sea level regimes, can be distinguished: (1) up to the beginning of the 1930s, (2) from 1931 to 1977, and (3) from 1978 to the present. The first of these periods — about 100 years — had a stable level, a mean annual level of about –26.2 m and a maximum amplitude of fluctuations that did not exceed 1.5 m. While there were occasions of sharp rises and drops in the level within a 6- to 8-year period, the level rose or declined by 40 to 70 cm, but there were no prolonged periods of rise or drop.

A sharp and prolonged decline in sea level was typical for the second period. Beginning in 1930, the level declined annually and, by 1941, the mean annual reduction in level was 16 cm. Further, the level continued to fall, changing only with increases in specific years. The total decline in Caspian Sea level for the second period was 2.8 m. In 1977, the sea level reached its lowest mark for the period of instrumental records.

In the third period (from 1978 to 1995), a continuous rise in level over 18 years averaged about 14 cm per year. The total rise in sea level in this period was 2.5 m. A major feature of this period is the intensification of economic activity with regard to the use of water resources in the sea's basin. Beginning in the 1950s, there were large-scale developments of irrigation in the river basins of the Volga, Ural, Terek, Kura, Sulak, and Sefid Ruda, among others. Many reservoirs were built on the rivers, industrial and municipal water use increased several times, and the water regime and water budget of the flood plains of the large rivers also changed. As a result, there was a major reduction of inflow of river water into the sea. If the above-mentioned hydraulic measures had not been carried out, the sea would have received even more water. Its surface area would have changed along with such climate characteristics as precipitation and evaporation. Outflow from the sea to the Kara-Bogaz-Gol Bay would also have increased. Taking into account all of these changes, special studies have allowed us to obtain characteristics of the sea level in the hypothetical absence of economic activity (see Fig. 1). Analysis shows that the period of rise would have begun not in 1978, but significantly earlier, in 1955-56, and that the sea level would not have fallen below about –28.0 m. These findings suggest that the total reduction in level in the 1970s would not have exceeded 1.5 m. And, in the absence of economic activity, the current level would be higher by about 0.7 to 0.8 m. In certain years in that period, the sea level could have been higher by 1.0 to 1.5 m, compared to its actual level.

The Reasons for Sea Level Fluctuations and Opportunities to Forecast Them

After the final separation of the Caspian Sea from the Black Sea basin, and the end of any hydrographic connection to it, the sea was transformed into a drainless water body — a lake. The water budget of such water bodies is wholly governed by climatic conditions within its catchment area. For annual and longer time intervals, the following formula can be used to determine the change in the volume of water in the sea:

$$x + y - z - u = \pm w$$

where x is precipitation over the sea's surface; y is surface and subsurface inflow to the sea; z is evaporation from the surface of the sea; u is flow into the Kara-Bogaz-Gol Bay, and w is the change in water volume in the sea.

For the geological period measured in millennia and longer time scales, there are possible changes in the basin form caused by geological processes: uplift and subsidence of certain areas, overthrust of platforms, volcanic activity, landslide processes, erosion, sedimentation, etc. Naturally, in such cases the geodetic marks of the level will have different values, even for similar elements of the water budget.

During the historical time period, measured in decades and centuries, the influence of such geological processes on the form of the sea basin has been relatively insignificant. Modern geophysics provide numerical assessments of these processes. Calculations of feasible changes in the volume of the basin yield negligible values and are well beyond the limits of measurement accuracy of any of the components of the sea's water budget. In this connection, hypotheses about the possible influences of geological processes on the present dynamics of sea level are not realistic.

The principal reasons for change in the water budget and, thus, in the sea level, are climatic. Annual changes of evaporation in the catchment area cause changes in the volume of river flow and in the amount of precipitation onto and evaporation from the sea's surface. The influx of groundwater, although still instrumentally unmeasured, most probably has the highest annual fluctuations, but its value does not exceed 1% on the "credit" side of the water budget.

Adequately detailed studies of the elements of the water budget are being carried out today, using available instrumental measurements. These studies convincingly show that the dynamics of the water budget correspond strictly to geophysical concepts about the hydrologic regime of inland (lake-type) water bodies.

Major studies have been carried out by scientists throughout the twentieth century to improve our understanding of the sea in order to forecast the dynamics (and level) of the sea for periods of one to several years, even decades, ahead. The most successful are the sea level forecasts made for one year in advance.

The problem of forecasting the sea's water budget and its level several years in advance can be resolved only by solving the global problem of climate forecasting for the same lengthy period. The Caspian Sea basin is rather large and occupies an area of about 3.5 million km^2. However, it takes up a little more than 2% of the land and less than 1% of the Earth's surface. Obviously, the water cycle in such a small area of the planet is ultimately determined by global climatic processes. However, the scientific community is still powerless to forecast those dynamics for a number of years ahead. Some climatologists believe that the problems of climate forecasting can be solved within the next decade or so, but not earlier.

Various widely used probabilistic methods for calculating sea level dynamics have been incorrectly used in a number of cases. The conclusions about sea level dynamics, made in the 1960s and 1970s on the basis of these methods, proved to be erroneous and led scientific researchers and practitioners astray. Now it is evident that these methods are not intended for real-time forecasting and cannot solve this problem. These methods are just a tool, a mathematical instrument, for the processing and analysis of initial information, but not a tool for providing the missing information. The main initial information needed for a data base for a water budget forecast should be, as a minimum, forecast data of river flow and of evaporation from the sea's surface. The inflow of river water can be forecast with a high probability for periods of less than one year. The data on precipitation and water accumulation in the river basin during the autumn to summer period provide a geophysical base; the latter is of major importance for determining the value of impending river flow.

There is no good comparable information for longer than one year. Therefore, such a reliable forecast today is not possible. Unfortunately, the value of the evaporation you can measure from the surface area of the sea cannot be forecast reliably even for one year in advance, because there is no corresponding reliable information for the same period on the amount of precipitation, evaporation, or other meteorological elements. This makes it difficult to quantify these parameters. However, the fluctuations between years in measured evaporation, although significant, are essentially less than the river flow and the input of their changes into the alteration of the volume of water in the sea (and its level) are also less. This makes it possible to obtain rather successful forecasts of the sea level for one year ahead, based on forecast information of the inflow of river water.

Thus, one should not draw pessimistic conclusions about the efforts to develop a reliable and credible sea level forecast. Studies providing new information on the changes of the values of evaporation and precipitation over the surface area of the sea are needed. Also needed are studies for the improvement of the reliability of annual forecasts of river flow and ways to extend the temporal scale of the forecast (up to two years). Finally, probabilistic calculations of the possible dynamics of sea level should be carried out using different forecasts or other information to identify various scenarios for the management of natural resources in the basin.

Sea Water Management Opportunities and Necessary Measures

The notion of "water regime" for a water body includes a wide range of geophysical, biological, medical, and economic aspects of an aquatic environment. In spite of the exceptionally huge size of the water body (with a volume of 78,000 km^3 and a surface area of about 400,000 km^2), several characteristics of the water regime have changed significantly within the recent decades, because of the impacts of economic activity. First of all, changes can be attributed to water pollution by such foreign matter as oil products, phenols, and metals. As a result of human intervention, important ecological conditions in the northern part of the sea were maintained, including the establishment of commercial sturgeon stocks. This points to the potential for opportunities for purposeful human influence on some aspects of the Caspian water regime and for national, as well as international, environmental management.

In the early part of the twentieth century, when the first plans of significant withdrawals of river flow for irrigation were developed and the possible reduction of water inflow into the sea was recognized, the possible decline in sea level and the reduction of commercial sturgeon stocks became a matter of great concern. This concern intensified when the sea level began to decline because of climatic changes and because of the reduction in river flow that began in the 1950s as a result of hydraulic construction.

Various proposals for the replenishment of water resources of the sea, including the reduction of evaporation from the sea's surface by separating the shallow water areas from the rest of the sea, the termination of outflow to the Kara-Bogaz-Gol Bay, connecting the Caspian with the Black Sea by way of a canal which would supply water to the Caspian, were put forward by many prominent Russian scientists and engineers. Some of these proposals were developed by planning and design organizations. Only one project, however, was realized, that of damming Caspian flow to Kara-Bogaz-Gol Bay. This was undertaken in 1980 at a time when there was no need to do it because the sea level had begun to rise quickly under the influence of changed climatic conditions. The bay was partially opened in 1982 to allow Caspian water to flow once again into the bay. Flow into the bay was fully restored in 1992. During the time when Caspian water outflow into the bay was limited, it saved about 160 km^3 of sea water. This contributed 40–45 cm to the sea level rise.

From the 1950s, hydraulic engineering operations for the construction of water reservoirs, the development of irrigation systems, and water withdrawals for municipal and industrial purposes were actively carried out in the basins of all major river systems that feed the Caspian Sea. As a result, the river inflow to the sea that could have been expected naturally was reduced. The reduction of flow for certain river systems reached as high as 70 to 80% of their natural flow. Volga River flow was reduced by 13 to 14%. Changes in river flow volumes were very significant and, when added together, they introduced a definite trend into the dynamics of sea water volume and level variations. Careful consideration of this factor demonstrated that the decline in sea level with no river flow withdrawals would have stopped twenty years earlier and not in 1978.

The last two factors indicate that by way of purposeful economic activity, there is a real possibility to affect, i.e., determine, the level of the sea. An assessment of technical developments made previously allows us to formulate the management problem in the following way: develop a system of engineering and management measures for reduction of the range of annual fluctuations in sea level. This system would be aimed at the prevention of possible damages from an expected disastrous rise or decline in Caspian Sea level in the forthcoming year.

The engineering measures should include the establishment of various hydraulic structures for river-flow redistribution and for the regulation of the outflow from the sea and of its surface area. Management measures should include economic policies to influence water consumption and water use in the river basins, policies to coordinate networks of reservoir operations, and other legislative measures. As a whole, it will require the establishment of integrated management systems for coordinating the actions of the countries in the Caspian region. Preliminary calculations show that within a year, the range of level fluctuations can change by 12 to 15 cm, or more.

A more radical solution to the problem is possible with the establishment of a system of regulation of the sea's water budget, one that includes the Azov, Caspian, and Aral seas. This scheme could assure a stable Caspian Sea level and the maintenance of minimum annual variations.

The Sea-Lake Problem: New in the Water Body Morphology

As noted before, the contemporary Caspian water budget and level regime are typical of inland lakes. The Caspian Sea is, in fact, the largest lake in the world, both in water volume and water surface area. It is a lacustrine type of hydrologic regime with its sharp interannual fluctuations in level. Such fluctuations have taken place throughout its existence as an enclosed water body. The lacustrine type of water budget and sea level is also witnessed in its intra-annual variations in sea level: with the beginning of spring flooding of the rivers, its level rises, sometimes up to 39 cm and with an increase in the intensity of summer evaporation, it is reduced back to its initial level and sometimes even lower. The lacustrine type of water regime is also manifested in its water salinity and thermal structures.

However, the Caspian water body also inherited certain features of the world ocean (e.g., its range of species of the animal world that are inherent to the ocean). The water body is linked to the sea by certain forms of current and wind regimes. Figuratively speaking, it is a sea in its water and a lake along the coast.

Until recently, when considering the geomorphology of the sea bottom, no data were available to determine shelf zones using a cartographic base, boundaries of debris cones of different rivers, or the position of paleo-channels of rivers. A new map of bottom topography was developed in 1994, using a method for mapping gravity flows (Fig. 2). Structures of the deltas of the Volga, Kura, ancient Amudarya, and other rivers of the basin can be seen on this map. The boundaries of the deltas and zones of sedimentation, formed by the rivers and adjacent territories of the different states, are clearly denoted for each basin. It provides a basis to demarcate objectively the boundaries of the bottom area gravitating toward certain coastal zones. It also allow us to solve problems related to identifying responsibility for the pollution of bottom sediments of certain coastal areas. Finally, it makes it possible to link certain sections of the sea bottom to specific littoral states.

The special value of this map is that it facilitates geophysical studies for the exploration of mineral resources. Based on specific features of the method of mapping and with regard to geologic features of the development of shelf zones, the map points directly to the location of possible deposits of mineral resources, especially oil and gas. The Uzboi delta (Turkmenistan), occupying nearly one-fifth of the sea bottom and having the shape of an arc, with a diameter of 250 km, is of great interest for those concerned with projecting the location of oil- and gas-bearing deposits. The delta is located in a deep oceanic depression, having no granite layer, and with a thick layer of sediment (up to 20 km) covered on top by mud volcanoes. The morphological similarity of this delta with the Euphrates delta, which contains the world's largest deposits of oil and gas, allows one to suggest by analogy that a large oil and gas field might be located here.

A series of oil and gas deposits might exist at points 3 to 7 (see Fig. 2). These are related to the zone of outcrop of productive strata of the delta tails A, B, and C. The most

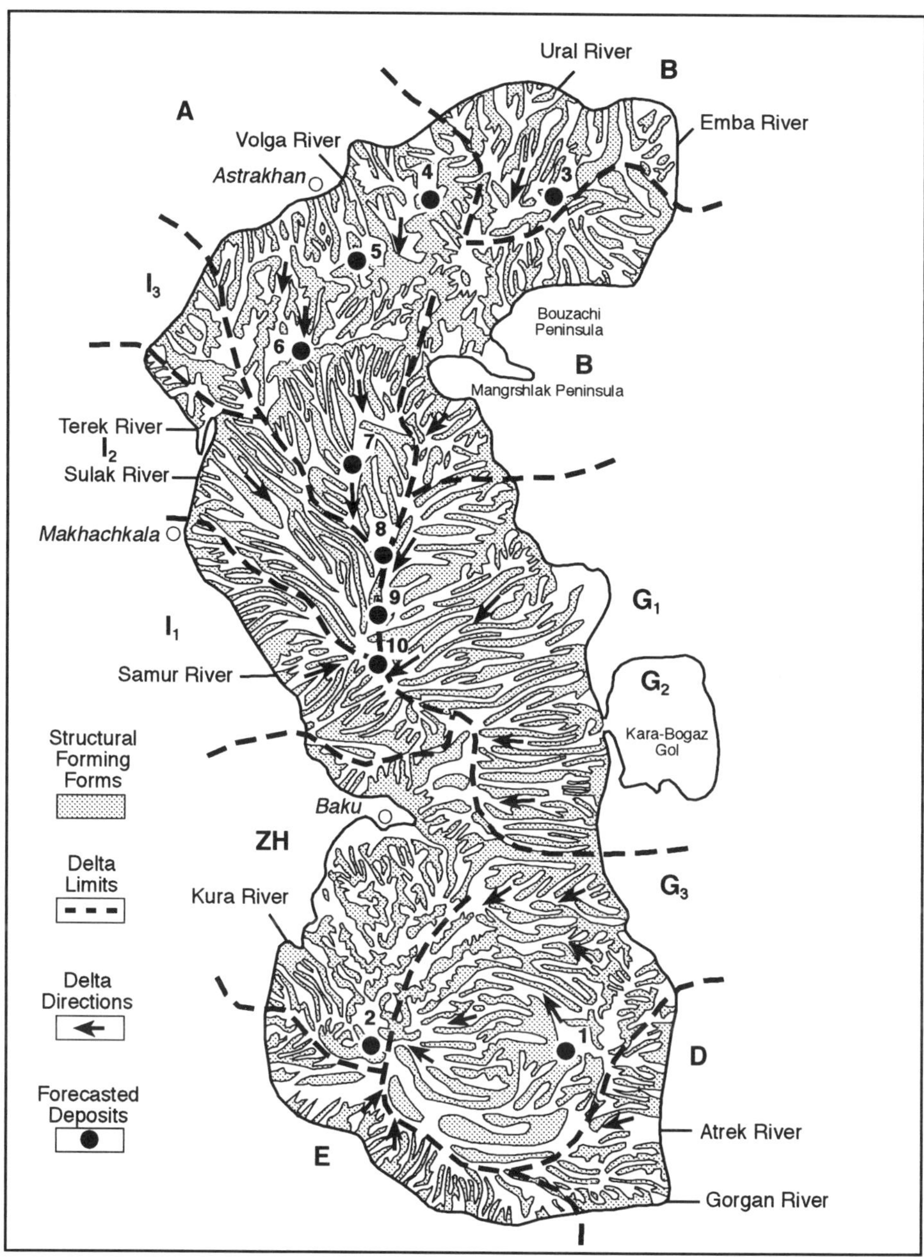

Fig. 2. Structural-morphological deltas building on Pliocene-Pleistocene surface of the Caspian Sea bottom. Scale 1:5.8 mil.

probabilistic large-scale deposits of oil in the region of the Volga Delta should be explored in the zones of points 8 and 9. Slightly to the south, in the region of maximum submergence of the bottom for this part of the sea (point 10), there is probably located one of the largest deposits of oil and gas in the Caspian Sea. For millions of years, this zone has served as an accumulator of organic matter, carried to the sea by the flows from vast areas of Europe and Asia.

Further study of the structural and morphological patterns of the sea bottom permit us to be more refined in our concepts about natural resources, and to facilitate not only their exploration, but also their joint use through coordinated regional decisionmaking.

Interstate Cooperation in the Caspian Sea Region

Until recently, water regime studies of the Caspian Sea had been carried out by various organizations of the former Soviet Union. Most of these organizations (and their scientific personnel) are now concentrated in Russia. All archives and data banks are also centralized in the Russian organizations. Taking into consideration the fact that the water regime of the sea is heavily influenced by flow from the Volga River and that the major portion of the biological resources in the Caspian are produced in Russian waters, studies on basic problems of the Caspian water regime and the use of its natural resources are actively being pursued in Russia and will likely increase in the near future. All states in the Caspian Sea region are interested in the data derived from these studies.

At the same time, Russian organizations are interested in multipurpose studies of the sea, encompassing all areas of the sea and coastal regions in its basin. Hence, the mutual interest of all states in the Caspian region should be to coordinate their scientific research efforts to establish a unified multipurpose program, which would be jointly coordinated and financed.

A program was developed in 1994 for an integrated international geophysical experiment in Russia to study the Caspian water regime (CASPEX). This experiment received the support of specialized organizations in Russia, Iran, Azerbaijan, and Turkmenistan. However, financial support for carrying out the program is still lacking.

In connection with the growing interest of many international organizations in oil- and gas-bearing resources of the sea, the demand for knowledge and forecasts on various aspects of the water regime of the sea has increased: sea level, dynamics of currents, wind surges of water along the coastal areas, transport and transformation of oil, organic and other pollutants in the aquatic environment, ice regime in the northern part of the Caspian Sea, the life cycle of fish stocks, and the ecology of the environment, and so forth.

In order to solve social and economic problems, there is a need for an assessment of natural resources of the coast and sea, and for the development of legal and juridical instruments for the management of the Caspian environment. Similar world experience is important for resolving analogous problems. Experience from around the world is also useful for obtaining new information about the sea and its basin. The development of integrated international programs is also necessary, while taking into account the national interests of the states in the Caspian Sea region.

In connection to this, it seems expedient, with the support of international organizations, to establish a Working Group made up of representatives of the states in the Caspian region, which should be charged with the preparation of a report for the governments of the states in the Caspian region on scientific issues to resolve social and economic problems related to the management of nature. The report should include a plan of action to implement top-priority studies, as well as the collection and analysis of information about the Caspian Sea region. The report should be distributed through the representatives of international organizations (e.g., UNESCO, UNEP, UNDP, World Bank, among others) in each littoral country. An international conference should be organized to include government officials, businessmen, and scientists.

In sum, what can we do immediately?

1. Determine whether the essence of the problem today is the unpredictability of the sea level or the inability to define the correct management of nature strategy under conditions of a changing Caspian Sea level.

2. Discuss the fundamental concept of the strategy for the rational management of nature, under conditions of a fluctuating and changing climate.

3. Putting aside for the moment comments on possible strategies, we highlight the necessity to organize research on and to discuss immediately the following problems:

    - We must obtain new multifaceted experimental geophysical information on the water regime of the sea;

    - The governments of the states in the Caspian region must create and adopt international legal documents on the reproduction, conservation, and use of natural resources; and

    - We must establish an international (e.g., regional) center for coordination of research studies and develop recommendations for the governments on nature management strategies.

PROBLEMS OF ECOLOGICAL SAFETY OF THE ASTRAKHAN REGION DUE TO THE RISE OF THE CASPIAN SEA

YURIY S. CHUIKOV
Committee of Ecology and Natural Resources
Astrakhan, Russia

The Caspian Sea is the largest enclosed water body in the world characterized by considerable fluctuations of its level. During the period of instrumental observations, fluctuations from —25.3 to —29 were detected. From 1837 to 1933 the sea level varied in the range of —25.3-26.6 meters. It was followed by a decrease of the level by 290 centimeters from 1933 to 1977.

Against the background of change of the Volga influx and the fluctuations of the Caspian Sea level, the events associated with intensive and sometimes unthought of and unbalanced economic activities of people in the Northern Caspian are unfolding.

It manifests itself the most in regulation of the Volga flow, which ended in the late 1960s and led to the considerable restructuring of the ecosystems of the Lower Volga and the Northern Caspian. The intensive pollution of the Volga and Caspian waters coincided in time with these processes. The reservoirs on the Volga turned into the accumulators of highly polluted bottom sediments, which on one hand somehow reduced the amount of pollutants coming into the Volga and the Caspian Sea, but on the other hand formed the basis of a large ecological crisis and catastrophes for future generations. They will have to resolve the problems of coping with colossal amounts of pollutants.

The economy of the Northern Caspian was historically oriented basically toward fish catching and processing, as well as toward agriculture which determined the development of the necessary infrastructure. This is why questions about the preservation and reproduction of fish resources and the condition of the lands play the most important role in the region.

However, during the past decades, due to the high anthropogenic load caused by the violation of nature conservation requirements, the ecosystem of the Caspian Sea suffered serious changes, represented by a reduction of its fishery

M.H. Glantz and I.S. Zonn (eds.),
Scientific, Environmental, and Political Issues in the Circum-Caspian Region, 119-129.
© 1997 *Kluwer Academic Publishers. Printed in the Netherlands.*

potential. More than that, in the Northern Caspian and in the Volga delta it led to an ecological crisis.

The following factors influence the ecological situation: emissions into the atmosphere of polluting gaseous and solid substances, discharges of polluted waste waters, violation of the requirements to the conservation of soils during development of deposits, pollution of the territory with garbage and industrial wastes, degradation of soils because of poor ecological practices in agriculture.

The lowering of the sea level led to the formation of vast shallow water areas, the appearance of drained islands, and macrophyte growth in the fore-delta. The construction of fish passage canals led to considerable changes in natural processes in this area. The vast shallow water areas turned out to be divided by the canals and artificial islands along them into separate sites.

Shallow areas, limited by artificial canals and shoals adjacent to the region of declining depth, gradually turned into enclosed water-bodies with distinctive vegetation and fauna of invertebrates and icthiofauna.

Intensive changes, caused by the lowering of the Caspian Sea level and by economic activities, were observed both in the lower part of the delta and in the fore-delta. The formation of vast shallow water areas, the appearance of newly exposed islands, and macrophyte growth in the fore-delta created favorable conditions for habitats of near water birds and waterfowl. This period was characterized by numerous birds nesting in these areas.

In accordance with the Ramsar (Iran) convention on wetlands of international importance, adopted in 1971, the USSR undertook the obligations aimed at the conservation of the Volga delta as a wetland of international importance. After the breakup of the USSR, Russia took over these obligations. At that time the territory of the Volga delta wetland was about 650,000 hectares.

In 1995 the Committee for Ecology and Natural Resources of the Astrakhan region made a survey and detailed description of boundaries of this site and worked out its statute. According to these data, the specified territory of the Volga delta wetland, including the state biosphere reserve "Astrakhanski," is about 800,000 hectares (situated in the coordinates 45 degrees 23 minutes to 46 degrees 27 minutes north latitude and 47 degrees 33 minutes to 49 degrees 27 minutes east longitude).

In 1978 the rise of the Caspian sea level began, with an average annual rise of 13 cm per year. The total sea level rise since then has been over 2 meters. The rise of the sea entailed the destruction of ecosystems, formed in the fore-delta, which manifested itself in considerable changes in plant communities and in the composition and the number of animals inhabiting the area.

But, in spite of the continuing rise in sea level, the most difficult and socially dangerous problem for the population of the Northern Caspian is the increase of aridity processes. In the Astrakhan region these processes are most pronounced in Kharablinsky, Narimanovsky, Krasnoyarsky, Enotayevsky and Limansky districts; Out of 2.6 million hectares of pastures, 76.9% are exposed to wind erosion.

Social, economic and political problems are connected with the use of pastures on the so-called "Black Lands." According to the data of the last geobotanic review, out of 228,000 hectares of pastures on the territory of the Black Lands, 168,000 hectares (74%) are heavily degraded. Their productivity has been reduced from 10-15 to 1-5 metric centners per hectare (1 centner equals 100 kg). Healthy, highly nutritional herbs in the vegetation cover have been replaced by harmful and poisonous species. For instance, in Narimanovsky and Limansky districts of the Astrakhan region, they occupy 16.4% of the pastures area.

The question of the trouble with the Black Land pastures was reviewed in 1980. The Counsel of Ministers of the Russian Federation adopted at that time Resolution No. 465 (Sept 20, 1980), "On Additional Measures to Better Black Lands and Kizlyar Pastures," which, however, was never enforced. This work started only in 1985 after the second Resolution (No. 239, dated June 3, 1985) and the decision of the Astrakhan Regional Executive Committee (No. 601) were adopted. In accordance with it, the Zenzely station, the forest and irrigation station, and the "Primorsky" state farm specializing in pasture-grass seed-growing were created.

In 1989 the Council of Ministers of the Russian Federation approved the general scheme of measures against desertification of the Black Lands and Kizlyar pastures. However, as a whole, the program of 1985-1990 aimed at the restoration of pastures on the Black Lands was not completed. The process of implementation of this program was reviewed by the Board of the State Committee for Nature of the Russian Federation in December 1989 and was considered unsatisfactory.

In 1990 the Council of Ministers of the Russian Federation adopted a new Resolution (No. 191) dealing with the same problem. The Astrakhan Regional Executive Committee made a decision "On Urgent Measures to Raise the Production of Feeding Grounds and the Restoration of Ecological Balance on the Black Lands and Kizlyar Pastures in 1991-1995."

This analysis showed that poor financing has been the main reason for not completing the state program on the Black Lands and for the Resolution of the Council of Ministers of the Russian Federation No. 191.

At the same time, practice shows that agro-forestry, irrigation, and land cover reclamation activities on the degraded pastures in the arid areas are not only an efficient method for arresting desertification processes but are also a realistic means for increasing the productivity of agricultural lands. According to the data obtained from the "Agrolesmeliostroy" Division and the Scientific and Research Institute of Agriculture, Forestry and Irrigation, these activities will result in an increase in pasture productivity from 1-5 to 20-25 centners per hectare of consumable mass. It is necessary to continue and expand these activities.

In 1994 a creative collective from the Committee of Ecology and Natural Resources of the Astrakhan region in the framework of the Federal program "Ecological Safety of Russia" made an assessment of the ecological situation in the Astrakhan region using criteria for exposing areas of ecological trouble. The results of this work testify that the condition of some lands in Kharabalinsky, Yenotayevsky, Krasnoyarsky and Narimanovsky districts of the Astrakhan region can be characterized as areas of ecological disaster. These lands, however, are not included in the range of activities of the program on Black Lands and Kizlyar pastures, because of their geographic location. All of this certifies that the development and implementation of a new federal program to combat aridification processes in the Russian Northern Caspian region is required.

The use of chemicals has become a serious factor in the functioning of natural ecosystems of the region. Between 1972 and 1994, 33,900 metric tons of pesticides of different classes were used in the region.

The maximum amount of pesticides were used in 1972, 1977 and 1978: 2589.6 tons, 2399.6 metric tons and 2485 metric tons, respectively. Between 1979 and 1986, the amount of pesticides used varied (1746 to 2126 metric tons). Since 1987, considerable and consecutive reduction of the amounts of chemical applied for vegetation protection began: in 1987 by 15.6%, in 1988 by 15.4%, in 1989 by 27.9%, in 1990 by 37.4%, in 1991 by 18.4%, in 1992 by 57.9% (annually, compared to the previous year). In 1994, when compared to 1986, the amount of pesticide use had been reduced by 19 times.

From 1978 to 1986 the area treated with pesticides varied from 309,000 to 365,000 hectares. Since 1987 a vivid reduction of this index began: in 1987 by 12.8%, in 1988 by 14.4%, in 1989 by 2.3%, in 1990 by 29.3%, in 1991 by 5.4%, in 1992 by 31.8%, as compared to the previous year.

The amounts of herbicides used varied among the districts of the region for rice culture and the production of vegetables and melons. For example, in 1989 in the Kamysiaksky and Chernoyarsky districts, the share of herbicides for rice production was about 90% of the total amount of their use in the region.

The analysis of data from other districts of the region where there is no rice cultivation, for instance in the Limansky and Narimanovsky districts, shows that the reduction of the amount of pesticide application and of the treated areas is due to reduction in the share of insecticides and fungicides — copper- and sulfur-containing preparations.

The pollution of water bodies in past years was also connected with direct input of pesticides into surface waters during aerial spraying of chemicals. In recent years these type of activities have been strictly limited and have been conducted using small amounts and in areas not connected to the water protection zones.

A serious problem is the pollution of agricultural soils on a scale of being able to cause background pollution of water sources due to the migration of chemical residue along the profile of groundwater and surface flow.

A toxicological study of agricultural lands undertaken 3-4 months after treatment with pesticides shows several cases where the toxic substance content in the soils exceeded allowable levels.

On the regional scale, during these years, in 52 samples the content of trace quantities of pesticides exceeded the maximum allowable concentration. The most unfavorable years within the study period were 1983 and 1987, and in the latter significant toxic residues were observed on 11 farms in 5 districts of the region.

Starting in 1988, no excess was observed of maximum allowable concentrations of pesticides in the soil of agricultural sites of the Astrakhan region. Trace quantities of pesticides within the limits of maximum allowable concentrations are found more frequently (9% of the cases). At the same time, beginning in 1987 there has been a tendency toward a decrease of toxic residue in the soil of farms in the region.

At the same time it was found that in the soils everywhere there is residue of both treflane- and chloride-containing compounds, GHCG and DDT, which have not been used in the region for many years.

Although the detected remainders of pesticides were significantly lower than the maximum allowable concentration levels, their presence on a large part of the territory and in different agro-climatic areas of the regions can serve as proof of the slow degradation of pesticides.

Even under present-day conditions of low chemical pollution levels in the agricultural lands, there exists one other problem connected with these pollutants. Intensive use and large amounts of chemicals brought into the region led to the appearance of local areas of highly polluted soils, including the areas subject to flooding resulting from the sea level rise.

In 1993-1995 the Committee for Ecology and Natural Resources of the Astrakhan region and the Astrakhan chemistry station conducted research on the storage yards of mineral fertilizers and pesticides in some of the districts of the Astrakhan region (Kamyziaksky, Volodarsky, Privolzhsky) situated in the water protection zone. A considerable part of these districts will become flooded with a further rise of the sea level.

Research showed that out of 39 warehouses only 18 were built according to a common design, but even they do not correspond to nature protection requirements. They are in poor condition, are not equipped with concrete slabs and drainage, fences are lacking, and so forth.

In spite of the fact that several years ago the Committee for Ecology and Natural Resources of the Astrakhan region had arranged for the collection and transportation beyond the limits of the region for burying or reprocessing banned and expired chemicals (over 100 metric tons), there are still considerable amounts of such substances in the warehouses. The lands around the warehouses are polluted with these remainders of pesticides and fertilizers; they are stored above the ground and without shelter. This has led to the high levels of local pollution of soils at a distance of 5 to 25 meters from the warehouses. What worries us the most is the accumulation of peritroids of dezis and karate, which have been used in the region only recently (since 1989).

Because of the flooding of some of the territory of the study area, it is necessary to take urgent measures to thoroughly inspect and decontaminate local pollution sources. Preliminary calculations show that in the process of liquidation of 16 warehouses and polluted sites, it would be necessary to render harmless about 1,500 to 2,000 cubic meters of soil.

Under the initiative of the Committee for Ecology and Natural Resources of the Astrakhan region, work started in 1994 on the realization of the program for an "Assessment of the Level of Chemical Pollution of the Territory of the Astrakhan Region." The work is being conducted by the laboratory of regional ecology of the analytical center of the State Institute of Non-Organic Chemistry of the Russian Academy of Sciences (work was done for the territory of the city of Astrakhan several years ago). The first results determining the places of local pollution based on several indices have been published. This research will allow us to develop a complete picture of chemical pollution in the areas subject to flooding and, later, for the whole region.

Many years of research have shown that a considerable negative influence on the water bodies of the region is caused by collector and drainage waters of irrigation systems from collective and state farms.

The Astrakhan Oblast is an area of considerable irrigated land use. It has its own methods for using irrigated lands. The majority of the methods were developed using an open drainage system to conduct drainage waters into water bodies of the flood plain and into the Volga delta. Because of this process, the use of pesticides, and the peculiarities of the water needs of such crops as rice, conditions are created for washing trace quantities of toxic substances into the drainage network and eventually into the water bodies.

The ten years of observations showed that the water bodies in the region were polluted mostly with rice herbicides: Saturn, propanid, jalan, among others. The maximum allowable concentrations (MAC) were exceeded on average and maximum indices of the content of pesticide residuals in the discharged water varied for jalan from 2.1 to 71 times, for Saturn from 2.5 to 100 times, and for propanid from 1.5 to 5 times.

Having assessed the concentrations of pesticides in soils, water sources and vegetative cover in the region, we can conclude that their content has diminished considerably during the past years.

Meanwhile, surface waters are still characterized by high pollution levels, which depend to a considerable extent on the quality of water coming from upstream. So, for instance, in 1992, as well as in the earlier years, on the border with the Volgograd region, the content of hydrocarbons in water was 3-4 times the MAC, for copper it was 4-5 times the MAC, and for iron 4-6 times the MAC. The pollution of Volga water increased after it passed the Astrakhan Industrial Complex. For example, hydrocarbon pollution in 1992 below the city limits reached 5 times the MAC, and copper 7 times the MAC. Pollution with hydrocarbons is connected primarily with water transport. The increases in the value of other indices are conditioned by the faulty operations of cleaning facilities.

Volga water quality influences the health of ecosystems of the Astrakhan state reserve. Observations conducted over several years have shown that the ecosystem of the reserve is subject to pollution with hydrocarbons, phenols, salts of heavy metals, etc., sometimes even to a larger extent than the areas of the Volga delta with its intensive economic activities. This is conditioned by the geographic location of the reserve sites in the lower area of the delta and in the fore-delta, which forms a certain trap for pollutants coming in with the water from upstream.

Considerable pollution of natural ecosystems in the region is caused by the emission of harmful substances into the atmosphere by industries of the Astrakhan Industrial Complex and the Astrakhan Gas Complex (AGC).

The total emissions into the atmosphere by the enterprises of the region has grown considerably in 1987-88. This has resulted from the start-up of the gas processing plant on the Aksaraisk deposit (the share of input of the "AstrakhanGasProm" in the total emission of the polluting substances into the atmosphere from stationary sources was 14.5% (1986), 47.3% (1987) and 88.1% (1988).

Later, as a result of the measures taken and the significant reduction of production (first of all on the AGC), total emissions of the polluting substances into the atmosphere were noticeably reduced and in 1992 reached the level of 1985. In 1993-94 an increase of emissions was observed as a result of the increase of production on the AGC.

Since 1988, inputs into the total amount of emissions into the atmosphere from mobile sources have been recorded, which is constantly growing both in absolute and relative values (in 1988 34.9%, in 1994 75.8% of the gross emissions into the atmosphere of the Astrakhan region).

The decrease in the number of sturgeon is creating a serious problem in the Volga-Caspian region. Without going into detail, let's review the change in sturgeon catches. The change is determined by such factors as the reduction of spawning grounds, excessive catches, water pollution, etc.

From 1901 to 1915, the catches of sturgeon in the Caspian were between 34,000-38,600 metric tons per year. The highest catch of Caspian sturgeon in the 20th century was 38,800 metric tons in 1913. After this over-catch, a gradual drop in catches began and, although fishing was carried out only in the sea. In the mid-1930s the catches were only 19,000-21,000 tons. Irrational, rapacious fishing for sturgeon in the open sea in the 1930s led to a drastic reduction in the number of fish, and in the 1940s the catch averaged about 15,000 metric tons.

In the 1950s, the average catch level was the lowest in the history of Caspian fishing (excluding the 1990s); it was only 11,400 metric tons, i.e., more than two-thirds less than in the first 15 years of the 20th century.

The 1964 ban on net fishing was the most important step toward the salvation of sturgeon from what appeared to be inevitable extinction. Already in 1971 the catch of sturgeon in the Caspian increased to 18,000 metric tons as opposed to 10,100 tons in 1960. By 1975 it reached 23,000 metric tons. In the following five years (1976-1980) the new over-catch started; the catches increased up to 27,300 tons in 1977 and averaged to 25,000 tons per year.

At this time a new powerful factor negatively influencing the number of sturgeon came in to action — the regulation of Volga flow. As a result of hydro-construction, the areas of the sturgeon spawning grounds was reduced by 7-10 times and made up to only 400 hectares in the lower, quiet reaches of the Volgograd dam. In the 1980s, the catches started to drop and in 1985-1988 they stabilized at the level of 14,500 metric tons in the Astrakhan region.

In the 1990s, the catches dropped catastrophically: in the Astrakhan region, 11,500 tons in 1991; 5,430 metric tons in 1992; 3,200 metric tons in 1993; and 1,400 metric tons in 1994. In 1995 the Caspian Scientific and Research Institute of

Fishery identified a sustainable limit of 1,500 metric tons. Hence, the potential threat of loss of production of these fish species was realized. The most considerable changes connected with the rise of the sea level have been observed in the lower part of the delta and in the fore-delta.

As a result of sea level rise, the shift of the sea edge of the delta occurred. The background sea level reached the lower parts of streams, going to the fore-delta, which led to an easy entrance of wind-induced surges into the delta.

The continuing increase of depth and the amplitude of water fluctuations caused by the wind-affected phenomena in the lower delta led to degradation and reduction of the areas of wildlife habitats as well as numbers of wild boar, fox, raccoon, muskrat and other animals.

The rise of the Caspian Sea level has also had a negative impact on the waterfowl population, causing the destruction of their nests, transformation of habitats, and the worsening of nesting and feeding qualities of their habitats. All of this added up to the lowering of productivity of the wetlands of the Volga delta, including the ones of international importance such as the waterfowl habitats. It also prompted the migration of some of the endemic animals into other natural ecological complexes (e.g.; the Ilmeny-hills district, Volga-Akhtuba flood plain).

The most highly productive area of the Caspian sea is its northern part and the Volga delta, where the basic mass of valuable fish species feeds and reproduces. Considering these conditions, the Government of the Russian Federation adopted in 1975 a resolution "On Declaring a Reserve Area in the Northern Part of the Caspian Sea." A similar decision was made in Kazakstan.

Regional economic activity provided for priority development of the fishery and water transport. At the same time the activities connected with prospecting and extracting mineral raw materials were banned.

In theory, the status of the reserve area was preserved until recent times. However, in Kazakstan the decision had already been made to start prospecting for and developing hydrocarbons directly in the wet areas of the Northern Caspian. More than that, according to information from the Institute of Oceanology (named after Shyrshov), in 1995-1997 the state organization "Kazakhstancaspyshelf" planned to cover the Northern and Middle Caspian with a network of seismic profiles with a total length of 35,000 km. According to the same source of information, by 10 June 1995, 500-km profiles were completed. Here, prospecting is being done by the fleet of small boats. The procedure includes the use of multi-channel bottom seismological cables, the special drilling of wells down to 10-20 meters, and every 12.5-25 meters with the explosion of small charges in them.

There have been some attempts to organize similar activities in the Russian zone of the Northern Caspian (in October 1995 one of the Russian companies got a license for the exploration for hydrocarbons in the Russian zone).

The use of shelf deposits of hydrocarbons can lead to the final degradation of the ecosystem of the Northern Caspian region and, as a result, will have negative socioeconomic consequences for fisheries in the Astrakhan region and in Russia as a whole. More than that, the uncoordinated activities of the states surrounding the Caspian Sea will, in the end, lead to unpredictable consequences for the ecosystem of the whole Caspian Sea, that is, the sea taken as a single whole natural complex.

At the first All-Russian Congress on Nature Conservation (June 1995) the draft project on the concept of transition of the Russian Federation to the sustainable development model was reviewed. In the preparation for the Congress on the Regional Conference for Nature Conservation, the measures for stabilization of the ecological situation in the Astrakhan region were identified.

Sustainable development of the state as a whole is not possible without sustainable development of its constituent parts — the regions. However, the conditions of development of each region can be completely different from one another. So, for example, the Astrakhan region is situated in the lower reaches of the Volga, which determines its natural conditions that provide for the existence of the population. On the one hand, hot continental climate, plenty of water from the Volga, and the highly productive portion in a biological sense of the Caspian Sea create favorable conditions for sustainable development of several branches of agriculture (cattle-breeding, fishery, plant-growing), of industry (fishing and fish processing, agricultural products processing) and of services (transport).

On the other hand, the cyclic fluctuations of the level of the Caspian Sea (which sometimes inundates and sometimes frees up considerable territories), the regulation and pollution of the Volga, an irrational undermining of fish resources of the Caspian Sea (which is taking place now as a result of the breakup of the USSR), attempts to divide the sea into zones of national interests, and the uncoordinated use of natural resources (including hydrocarbons), all bring regional development into considerable imbalance.

To improve the economic and ecological situations in the region it is necessary to change completely the strategy of nature use. It is necessary to return to the former priorities — the protection, in the first place, of the natural resources that are able to renew themselves. To this end priority should be given to fishing and traditional agriculture as the most ecological sustainable ones. It is also necessary to develop transport and trade. Ecological tourism should become a new goal.

The development of oil, gas and chemical industries should continue, dependent upon the complete elimination of serious adverse ecological impacts on the region's traditional economic activities. This is the only way to realize the strategy of sustainable development and the survival of the Northern Caspian.

BIOLOGICAL ASPECTS OF CASPIAN SEA LEVEL RISE

ZALIBEK G. ZALIBEKOV
PreCaspian Institute of Bioresources
Makhachkala, Republic of Dagestan, Russia

The Caspian Sea, the world's largest landlocked water body, is shared by Russia, Kazakstan, Turkmenistan, Azerbaijan, and Iran. It is characterized by significant and quasiperiodic millennial, centennial, decadal, and interannual fluctuations of sea level. Those fluctuations have had a maximum amplitude up to 25 m in the last 10,000 years, and up to 15 m in the last 2,500 years within a range of 25-35 m of the absolute land surface level. Six large-scale changes of the Caspian sea level with an amplitude from 5 to 10 m have been identified in the modern era, that is, the past two millennia. The most recent changes in level led to the deterioration of fertile coastal soils and have contributed to the replacement of highly productive delta alluvial landscapes by a marine (shoal) water surface (Fig. 1).

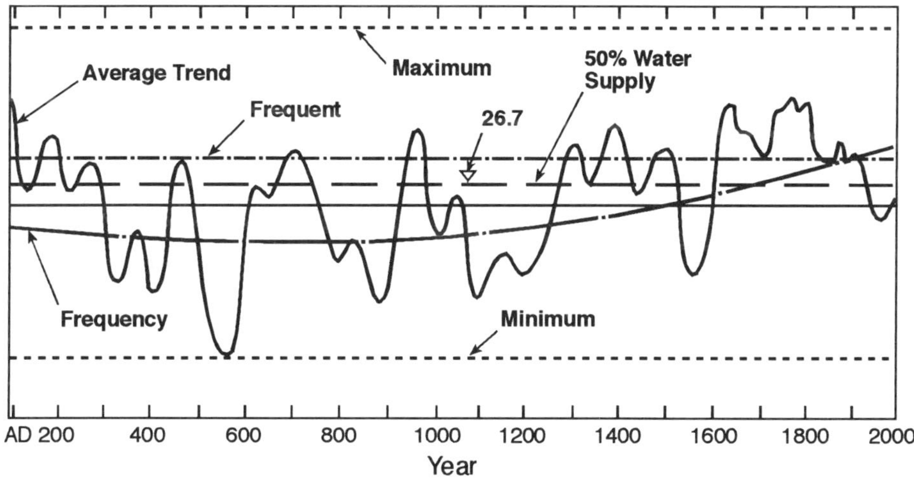

Fig. 1. Six large-scale changes of Caspian Sea level

Variations in sea level occur as a result of changes in atmospheric temperature and precipitation. A most favorable period with respect to the use of coastal lands lasted from 1837 to 1977, when sea level declined from −25.3 m in 1837 to −29 m in 1977. This period was marked by a movement of practically all kinds of economic activities

M.H. Glantz and I.S. Zonn (eds.),
Scientific, Environmental, and Political Issues in the Circum-Caspian Region, 131-137.
© 1997 *Kluwer Academic Publishers. Printed in the Netherlands.*

of society toward the receding shoreline. It was accompanied by the creation of highly productive agricultural lands and new settlements, and the formation of distant pastures, industry, and recreation grounds [1, 19]. Climate-related conditions in this period were characterized by a rapid drying of the sea bottom, ecological transformation of wetland and meadow communities to meadow-steppe and steppe and finally toward desert communities. A succession of transitional areas was formed in the coastal zone varying from highly productive meadows to low-productivity desert communities. A unique feature of this environmental change was the evolution of aquatic (hydromorphic) landscapes into terrestrial ones with progressive aridity, steppe formation, and desertification processes. Thus, under an increasingly hot and arid climate, aquatic ecosystems and their components transformed to terrestrial ones with subsequent xerophytization [8-10].

Caspian Sea level rise started in 1978 at an average rate of 13 cm per year and its increase is currently estimated at about 2.5 meters (−26.7 m). A rising trend continues today and according to some projections, sea level could probably reach −25 m in the next 10-15 years [10, 19] (Fig. 2).

Over 500,000 hectares of valuable lands have been flooded in the last 18 years and were withdrawn from agricultural use on the Caspian seacoast [10]. As of early 1995, the total environmental damage from sea level rise was estimated at over 4.5 billion Rubles (in 1991 prices). Located within the destructive zone of the Caspian Sea are the cities of Makhachkala, Derbent, Kaspiysk, Lagan, the village of Sulak, and numerous settlements and economic sites (buildings, infrastructure). It is noteworthy that construction of large-scale economic projects within the coastal zone continued even during the early period of rapid sea level rise from 1978 to 1985.

The intensive use by society of the coastal areas has resulted in a high density of housing, economic and recreational complexes, exerting pressures on land that had become exposed for 140 years. There are various opinions on the causes of Caspian Sea level rise [2]. Today, the most common view is that the water budget has been altered by an increase in the volume of annual runoff and a decrease in evaporation. Impacts on sea level from geodynamic, biodynamic, and anthropogenic factors were considered to be of low probability. It is useful, however, to focus attention on the anthropogenic factors believed to have a marked impact on Caspian Sea level changes. These factors are related to the biological potential of soils and their formation on porous modern Caspian sediments [4, 5, 7, 12].

We have formulated an hypothesis on the physical loads of coastal technogenic (i.e., human-built) landscapes on the vertical composition and compactness of geological deposits. Considerable technogenic loads on both the Caspian depression and other delta regions [8, 16, 17] result from a high biological productivity of soils and intense industrial, agricultural, and urban activities. Physical loads, during a vertical compression of geologic layers due to technogenic impacts, are shifted horizontally toward the sea bottom. Horizontal loads on the sea bottom

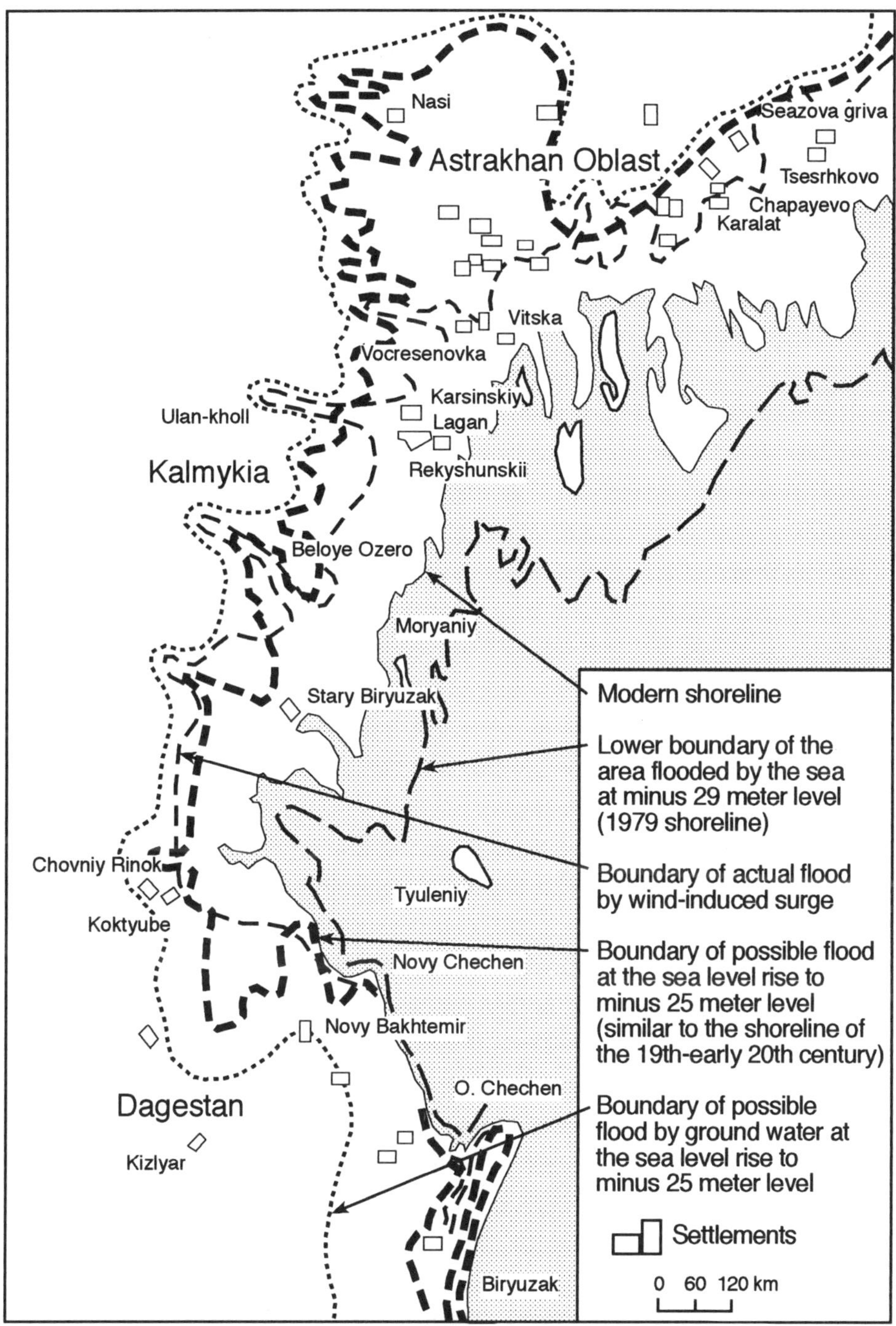

Fig. 2. Caspian Sea level rise

are, to a certain extent, stabilized by 40 million metric tons of suspended matter deposited annually by the Volga, Ural, Terek, Sulak, and other rivers.

Subsiding, stabilizing, and rising sea level processes are related to various factors. We believe the most important of these factors is the additional vertical loads of technogenic landscapes in the coastal zone. Technogenic objects, e.g., multi-story buildings, roadways, etc., account for 15% of the total coastal areas of Azerbaijan, Dagestan, Kalmykia, the Astrakhan region, Kazakstan, and Turkmenistan. The technogenic parameters measured by physical weight and load per square unit has increased by more than four times in the past 20-25 years.

The Caspian Sea basin is made up of modern geological deposits with ongoing subsiding, rising and contracting, processes. Over time they interact with the upward-rising surface, though with less stability [11, 18]. The tendency of the constituent materials to compression and the lack of stability in deep-lying layers contribute to a slow subsidence of the coastal plains. One can assume that the vertical compression of the layers and the horizontal shift of the load from the surrounding area are the results of an expanding sea-bottom surface and partial linkage of underground water with sea water.

Changes in coastal elevation are also related to an increase in the area of a zone of geological deposits and its reduced ability to resist vertical technogenic loads. The lower stability of the existing moistened geological layers contributes to a dynamic process of expansion and contraction, predominantly in the vertical direction, and facilitates the sagging phenomenon.

In light of the above, we have proposed a new approach to the Caspian Sea problem and formulated an hypothesis about a significant contribution to the sea level rise process of physical loads by technogenic landscapes.

To investigate the situation, we carried out ground mapping of the density of the technogenic cover and the dynamics over time of coastal elevation. Designing construction projects and implementing coastal protection measures should take into account data on sea level changes, as well as on the dynamics of rising and subsiding phenomena in the coastal plains. International organizations and circum-Caspian states should be encouraged to financially support the creation of an observation network to monitor seasonal, annual, and long-term dynamics of ground surface changes in the cities of Makhachkala, Derbent, Astrakhan, Lagan, the village of Sulak, and the administrative regions in the coastal zone. The data obtained would help us to identify the dynamics and interrelationships between changes in sea level and changes in the coastal zone.

Today we are witnessing a paradoxical situation: enormous budgets are allocated for coastal zone protection, with little or no data about the dynamics of the elevation required for the protection of buildings or other infrastructure. Research is needed to clarify the interrelationships among geological, climatic, and biological factors in Caspian Sea level processes leading to the accumulation in bottom sediments of suspended matter transported by rivers [15].

The environmental situation in most regions of the Caspian depression is characterized by two very different phenomena — flooding and desertification. They exhibit the following features:

- shrinking areas of terrestrial ecosystems, including arable lands;
- increasing anthropogenic activities in the region have led to the deterioration of soil and the vegetative cover and to the accelerated soil deflation and salinization processes;
- expansion of the surface area of the Caspian shoal, waterlogged and meadow soils, the formation of new soils, and the transformation from xerophytic conditions to the hydrophilic dryland conditions.

The changing hydrographic pattern in the region and the development of hydromorphism (e.g., an increasing water content) modified dryland processes. A specific feature is the local hydrophytization of automorphic soils (with no account for precipitation) that has become widespread in the coastal areas of the Caspian depression.

Inundated soils lose contact with the atmosphere and interact with the lithosphere and hydrosphere. Submerged soils or bottom sediments formed in the contact zone have different parameters and functions [6]. Cyclic recurrence of terrestrial and submarine features of soil formation and wind-induced surges modify soil morphology, although the soil horizon profile remains unchanged. The features of relief, vegetation, and constituent material also remain the same. The newly formed soils and their evolution are poorly studied, because of the unusual combination of terrestrial and submarine soil formation features. A hypothetical notion of "hydrosoils" had been introduced to define a focus of studies on seasonally flooded areas.

A shift of terrestrial complexes on the exposed, unflooded parts of the region suggests an indirect impact of the rising Caspian Sea level on human-induced desertification [10, 14]. The main type of changed activity is the grazing of 2,500 domestic cattle on 1,000 ha of pastureland. The grazing load on the unflooded part of the area increased twofold, and the density of economic activities 1.5 times, followed by a reduction in soil area, an increase in soil destruction (e.g., compaction and trampling), and a reduced protective cover of vegetation. Since the 1970s, the technogenic load has increased, as a result of the construction of roadways, large-scale cattle-breeding farms, housing, and recreational areas. As a result, over 100,000 hectares have been withdrawn from use because of sharply reduced biological productivity.

The high density of technogenic development in the unflooded parts of the region contributes to:

- lower productivity of existing soils, rapid xerotization, and desertification processes due to secondary salinization, deflation, and desiccation of soils and the degradation of pasture vegetation;

- increased vertically oriented physical loads causing changes in coastal zone elevation.

Growing technogenic loads have zero impact on the total land mass on a global scale. However, we believe the high density of technogenic activities in the coastal zone has generated major changes in land-use patterns and in human activities. The accumulation and concentration of the transformed land mass and biogenic energy occur mostly in areas of high biological productivity, e.g., coastal areas, rivers, deltas, alluvial plains, tidal terraces, and valleys.

The most fruitful Caspian research will likely focus on the reasons for sea level fluctuations from the perspectives of various fundamental sciences: biology, ecology, geology, climatology, hydrology, policy, and soil science. A multidisciplinary approach is necessitated by the complex nature and diverse direct and indirect aspects of the rising level of the Caspian Sea — the world's largest inland water basin.

References

1. Dagestan, ASSR (Autonomous Soviet Socialist Republic), 1979: *Agroclimatic Reference Book*. Leningrad: Gidrometeoizdat. (In Russian)

2. Berg, L.S., 1936: The Caspian sea level in history. *Problems of Physical Geography*. Moscow, T.P. (In Russian)

3. Vernadsky, V.I., 1940: *A Few Words on the Noosphere: Achievements of Modern Biology*. Moscow: Nauka Publishing House. (In Russian)

4. Golubyatnikov, V.I., 1949: Geological composition of Dagestan. *Proceedings of the First Session of Dagestan Branch of USSR Academy of Sciences*. Makhachkala. (In Russian)

5. Dobrovolsky, G.V., K.N. Fedorov, and A.V. Stasyuk, 1975: *Geochemistry, Land Reclamation and Genesis of Soils in the Terek Delta*. Moscow: MGU Publishing House. (In Russian)

6. Dobrovolsky, G.V., and B.D. Nikitin, 1989: *Soil Function in the Biosphere and Ecosystems*. Moscow: Nauka Publishing House. (In Russian)

7. Egorov, V.V., 1959: *Soil Formation under Irrigation in the Deltas of the Aral-Caspian Lowland.* Moscow: Nauka Publishing House. (In Russian)

8. Zalibekov, Z.G., 1980: Analysis of anthropogenic use of soil resources in Dagestan. *Soil Science*, **7**. (In Russian)

9. Zalibekov, Z.G., and A.G. Otsokov, 1993: Regression analysis of data on soil salinity in the Terek delta. *Soil Science*, **11**. (In Russian)

10. Zalibekov, Z.G., 1995: Experience of environmental analysis of soil cover in Dagestan. Makhachkala: Dagestan Scientific Center. (In Russian)

11. Zenkovich, V.G., 1968: *Principles of Coastal Development*. Moscow: Nauka Publishing House. (In Russian)

12. Zonn, S.V., 1993: Classification and geography of soils in the Terek basin. *Proceedings of LOVIUA*. (In Russian)

13. Zonn, S.V., 1995: Desertification of natural resources and control measures in Kalmyk agriculture in the last 70 years. In: *Kalmyk Biota and Environment*, I.S. Zonn and V.M. Neronov (Eds.), 19-52. Moscow: Russian National Committee for UNEP (UNEPCOM). (In Russian)

14. Zonn, I.S., 1995: Republic of Kalmykia: European region of environmental tension. In: *Kalmyk Biota and Environment*, I.S. Zonn and V.M. Neronov (Eds.), 6-18. Moscow: Russian National Committee for UNEP (UNEPCOM). (In Russian)

15. Klenova, M.V., 1958: *Marine Geology*. Moscow: Nauka. (In Russian)

16. Kovda, V.A., 1972: *The Biosphere and Humankind*. Moscow: Nauka Publishing House. (In Russian)

17. Kovda, V.A., 1974: Biosphere, soils, and their use. Report at the Tenth International Soils Science Congress. *Soil Science*. (In Russian)

18. Mozhaev, B.N., 1995: Impact of modern shifts in the earth's crust on desertification intensity. *Arid Ecosystems*, **1**. (In Russian)

19. Russian Federation Security Council, 1995: *Environmental Safety of Russia*. Moscow: Russian Federation, Issue 1. (In Russian)

III. Desertification

EUROPE'S FIRST DESERT*

TATYANA A. SAIKO
University of Plymouth
United Kingdom

IGOR S. ZONN
Russian National Committee
for UNEP (UNEPCOM)
Moscow, Russia

Europe now boasts its own desert: not one caused by its climate, but a dismal barren landscape created by the actions of humans. This anthropogenic phenomenon, covering a million hectares of once fertile pasture lands, lies in the Republic of Kalmykia on the northwest coast of the Caspian Sea.

Kalmykia has a population of just 350,000 in an area of the size of Austria, and is one of the 21 independent republics within the Russian Federation. A poor agrarian republic, its two main products — wool and oil — do not provide sufficient wealth for its people. Some 200,000 tonnes of oil is produced there annually, and each sheep yields only 2.7 kilograms of low quality wool. Erosion of the steppes, inherently fragile because of their high-risk soils and regional climate, has left the people unable to provide for either their animals or themselves.

But this was not always the case. In 1803, the steppes fed 2.5 million head of cattle, sheep, horses and camels, as well as a million or so saiga antelope. Indeed, before the 1917 revolution, Kalmykia was the "animal farm" for the whole of Russia, as well as supplying meat and wool to Persia, and Khiva and Bukhara to the east of the Caspian Sea. The ecological balance and fertility of the pastures were maintained for centuries by following a transhumance system. In winter, animals were taken to the so-called Black Lands (Chernye Zemli) - high steppe grasslands which remained green all year round. In summer, however, these areas were left alone and no grazing was allowed. While climatic fluctuations caused the nomadic economy to prosper and contract, the Kalmyk steppes were still rich with diverse vegetation and animal life in the 1920s. In 1913, 300,000 people lived in Kalmykia, while its

*This article first appeared in ***The Geographical Magazine*** of the Royal Geographical Society in Great Britain (Volume 67, No. 4, 1995, pp. 24-26). It is reprinted herein with their kind permission.

M.H. Glantz and I.S. Zonn (eds.),
Scientific, Environmental, and Political Issues in the Circum-Caspian Region, 141-144.
© 1997 *Printed in the Netherlands.*

animal population included one million head of sheep, 300,000 head of cattle, 200,000 horses and 20,000 camels. Sheep meat production was between 25 kilograms and 40 kilograms per head.

During the Soviet period though, the Kalmyks, like other nomadic races, were forced to establish a settled way of life. The creation of large collective state farms was accompanied by an increase in herd numbers and the disruption of rotation systems. Fat-tail sheep with flat hooves were replaced by the sharp-hooved merino and Astrakhan varieties which destroyed the grass and thin soil cover. Desertification in the steppes had begun.

A Series of Disasters

The first natural disaster came with the revolution and the civil war, when Kalmyks supported "white" Cossacks and were mercilessly punished by the Bolsheviks. The population fell to 135,000 by 1926. The second disaster followed in 1943 when, under alleged collaboration with the Nazis, 90,000 Kalmyks were deported to Siberia and Kazakstan. The human death toll was almost 50,000. As the indigenous population was expelled, traditional land management and conservation methods disappeared too.

Between 1943 and 1957, the territory itself was eliminated and became Astrakhan Oblast (province). But in 1957, the USSR Supreme Soviet issued a decree which established the Kalmyk Autonomous Oblast, and by 1958, Kalmykia had become an autonomous republic. In 1989, 45.5 percent of the population were Kalmyks and 37.7 percent Russian.

During the 1960s, Nikita Krushchev initiated a large-scale national campaign so that agricultural production would surpass US levels. In 1970, the sheep population was 2.3 million, and by 1985 it had grown to 3.4 million — outnumbering humans by 10 to one. At the same time, animal productivity declined sharply from over 25 kilograms of meat per sheep. This dramatic drop meant that the total meat production from three million sheep could have been achieved from just one million sheep in 1913.

The estimated carrying capacity of the Black Lands is less than 750,000 head of sheep, while current sheep numbers exceed 1.6 million. There is additional pressure on resources from cattle, which in 1990 numbered 350,000 head. The former grazing area of saiga antelope has fallen from 100,000 hectares to just 23,000 hectares, and poaching of these endangered animals is a problem. Today, total animal pressure on pastures in the steppe zone is nearly five times higher than the permissible level. Trampling of these overgrazed pastures has accelerated the effects of wind and water erosion.

The cultivation of lands for fodder and cereal production also contributed to the steppe's degradation. Barchan dunes began to form, and by 1970 the area of shifting sands reached 250,000 hectares; by 1980, it covered 500,000 hectares. The total area of desertified lands has increased dramatically from five percent of pasturelands in 1957, to 37 percent in 1970 and 83 percent in 1990. Today, an anthropogenic desert of one million hectares — three percent of the republic's territory — has been created in the Black Lands. Every year, the desert grows by about 50,000 hectares and it is estimated that if current land management practices persist, it will reach three million hectares by the year 2030.

However, there are some positive signs. Several desertification control measures are underway. But while a "Plan of Action to Combat Desertification in Kalmykia" has been implemented, any progress is dependent on the republic's economy, and this is in a desperate state. The average monthly salary is less than US$25, while collective farmers get a fifth of this and are forced to sell their animals to buy sugar, butter and salt. Since November 1994, the government has failed to pay even these meager salaries, thanks to 14 billion roubles mysteriously 'disappearing' in Moscow, probably through corruption.

Overall production fell by more than 30 percent in 1994, with processing industries' output falling by 40 percent and construction by 50 percent. Agriculture yields fell by a quarter.

The Political Will?

This year (1995), independent Kalmykia celebrates the second anniversary of the election of its first president, and Europe's youngest, 33-year old Kirsan Ilyumzhinov. A billionaire former businessman, Ilyumzhinov has gained substantial publicity for his populist and controversial policies.

He succeeded in peacefully dissolving the Supreme Soviet in the Republic of Kalmykia in May 1993, and introduces a tax system favorable to local enterprises and other measures to support the market economy. However, many of his pre-electoral promises are far from having being implemented. His plan to raise per capita income to $100 a month now seems unattainable.

He said he wanted to transform Kalmykia into a "second Kuwait" and a "mini-model for Russia," and promised to make it a self-sustaining republic, financially independent of Russia. In an attempt to achieve these targets, Ilyumzhinov established dictatorial powers, including the right to dissolve parliament. In April 1994, however, he abolished the constitution and declared that Kalmykia had no national interests of its own and was inseparable from Russia. Russian subsidies for production have increased from 80 to 90 percent (before Ilyumzhinov's presidentship), to almost 100 percent.

Without cash to spend, however, it is difficult to address the problem of Kalmykia's Black Lands — what projects are already in place seem to make little impact on the growing desert. In 1993, sheep numbers fell to 1.5 million but this has not noticeably improved the ecological situation. The area was declared an ecological disaster zone in 1991, and it is thought by many to be second only to the Chernobyl catastrophe in terms of its environmental impact.

On 1 August 1993, Ilyumzhinov declared a state of ecological emergency in the republic. But it remains to be seen whether he can reverse the environmental damage inherited from the totalitarian regime.

STUDIES OF AEROSOLS FROM KALMYKIA'S BLACK LANDS AND THEIR INFLUENCE ON ENVIRONMENT

I.G. GRANBERG
G.S. GOLITSYN
G.I. GORCHAKOV
E.M. DOBRYSHMAN
V.M. PONOMAREV
Institute of Atmospheric Physics
Russian Academy of Science (RAS)
Moscow, Russia

A.V. ANDRONOVA
Karpov Institute of Physical Chemistry
Moscow, Russia

P.O. SHISHKOV
Moscow Academy of Device Construction and Informatics
Moscow, Russia

B.V. VINOGRADOV
Severtzov Institute of Ecology and Evolution RAS
Moscow, Russia

Introduction

The influence on environmental quality of various pollutants discharged into the atmosphere, along with global climate change and changes in human health, sparked an increasing interest in the sources and transport of pollutants. Dust aerosols are one of the main pollutants of the atmosphere that influence the physical characteristics of the air and affect agriculture and human health. Today, there are regions in Russia with serious soil damage due to overgrazing (Kalmykia), and with drying up of wetlands (the Volga Basin, the southern Urals, etc.). A similar situation exists at Owens Lake, California, which is the largest source of airborne dust and salt in the US southwest. A similar situation exists in the Aral Sea region in the former Soviet Central Asia. These are regions with frequent dust storms and are major sources of atmospheric dust and salt. The dust storms further damage soils, vegetation and ecosystems. Similar processes are observed in many other parts of the world, especially in developing countries.

M.H. Glantz and I.S. Zonn (eds.),
Scientific, Environmental, and Political Issues in the Circum-Caspian Region, 145-153.
© 1997 *Kluwer Academic Publishers. Printed in the Netherlands.*

This chapter presents a discussion of the situation in the Kalmykia Republic (Russia), where the largest semi-desert in Europe is located, e.g., Kalmykian Black Lands. The Black Lands extend between the Caspian Sea on the east and the Ergeni highlands on the west and encompass an area of about 156,000 square kilometers. A high level of instability in semi-desert ecosystems on the one hand, and severe anthropogenic impacts on the other, have led to the creation of one of the world's largest regions of very severe desertification and ecological disaster. In our view this region is a regular source of dust and salt emissions to the atmosphere. The aim of our research is to investigate (1) the conditions and mechanisms for dust and salt transport from newly desertified territories, (2) the chemical and physical properties of the aerosols and soils and (3) to develop a model of dust and salt transport by wind from source to deposition areas. This study will provide a basis for estimates of the influence of these processes on local climate and on the environment. The preliminary experimental activities were carried out in July 1995.

Physical Peculiarities of the Region

The Geographical Features

The Republic of Kalmykia is located between 45° and 48°N and between 43° and 47°E. Its territory looks like a rectangle elongated from southeast to northwest. The longest diagonal is about 400 km. The middle cross-section from southwest to northeast is about 100 km. Kalmykia's coastline on the Caspian Sea to the southeast is about 120 km long. The southeastern part of Kalmykia is below sea level. The general landscape consists of some hills with a maximum height of 200-220 m. There are no large forests, but small areas of woods do exist in some places in the northern part of the republic. Numerous small lakes become dry in the summertime. So, a considerable part of the soil has a sand-salt composition.

Within the wave-like relief, three main kinds of soils can be identified: accumulative, deflation and mixed accumulative-deflation. The accumulation of eolian relief forms varies depending on vegetative cover, mechanical composition, relative height, and relief-forming winds. Drifting sands are divided by relative height as follows: 3-5 m and 1-2 m barchan ridges. They stretch from the north-northeast to the south-southwest and have asymmetric slopes. The semi-fixed sand forms have hillock relief with aggregations of large grasses, semi-shrubs and shrubs. Fixed sands have a gentle undulating relief. One can recognize old and young sand massifs. Old bare sand massifs occupy less than 5% of the total land surface of Kalmykia. They have less silt particles, but have heavy minerals and water-soluble salts transported by winds in previous decades and centuries. Most of the drifting sands are young, having originated during the last few decades (1960s-1980s). They have a high content of heavy minerals, small particles, soluble salts and sulfates. Deflated eolian relief varies depending on its relative age, size and depth, vegetation composition, soil moisture and salinity. Old deflated depressions cover relatively large areas (100-1000 m) and are associated with old wells. The depth of old depressions varies from 3 to 10 m and is limited by the level of groundwater, wet subsoil, clay layers, or dense vegetation. The old sand massifs are situated near old deflation depressions that originated on the overgrazed rangelands and resulted from the destruction of the vegetation and soil cover over the last few decades. Their depth is less than 3 m and encompasses an area of 100-1000 m. Finally, the deflation scarps from the 1960s have the greatest size of 1000-10,000 m and small depth of up to 3 m due to erosion. Marine deposits have become exposed on the bottom of these deflation scarps. These deposits slow down revegetation processes and hinder land reclamation. Young deflation scarps provide the largest source of dust to the atmosphere.

CLIMATE

Climate of the region is subarid and subcontinental, with hot summers and cool winters. The mean annual temperature is near 9.5°C; the mean for January is about −3.6°C; the mean for June is 25.5°C; the absolute minimum is −42°C and the maximum is 34°C. The mean annual precipitation is about 200-250 mm, varying from 300-400 mm in wet years to 100-200 mm in dry years. Southwest winds prevail, especially during the warm period of April to October.

The highest frequency of dry and dusty days is in August. Dust storms begin to occur in April, after the drying of the soils but before vegetation growth. They are more frequent in the July-August time frame, after the drying of the vegetation and before the autumn rains. The diurnal temperature amplitude may reach 23-25°C in clear weather. When atmospheric pollution is high, the diurnal temperature amplitude is reduced by a factor of 1.5 to 2. The amplitude of soil temperatures can be more than 45-50°C. For the greater part of the year, the land is in the west periphery of a weak anticyclone (high pressure system). As a result, weak easterly winds are generated. Moisture-laden winds from the Mediterranean and Black Sea lose their water content before reaching the west side of the Volga region. During the warm period, after the passage of decaying atmospheric fronts from the north or northwest, even weak winds can create rather intensive sand and dust fluxes. 4-6 days during the summer months, there is a high probability of a dust storm or/and heavy pollution transport into the region. In such days the typical synoptic situation is anticyclonic. The probability of strong winds is about 25-30% (with a velocity greater than 8-10 m/s), when mass transport becomes intense at midday in July. The probability of a dust storm is about 5%. The main direction of pollution transport is to the west; transport to the east is much smaller. To investigate the sand-soil transport in Kalmykia in detail, it is useful to identify a specific test site. Taking into account the spatial distribution of sandy areas and the dominant winds, the area bounded by 45°50' - 46°10' N and 46°00' - 46°50' E seemed favorable as a test site.

GEOLOGY AND SOILS

The test site is covered by Upper Quaternary silt sand marine deposits with clay layers (Qm hv3). [This is the international nomenclature for Quaternary marine hvalyn deposits.] They are changed by eolian processes and overlaid by eluvial and drifting sands. They are underlain by Lower Quaternary silt/clay marine deposits (Qm hv1-2), which are exposed on Pre-ergenian lowland, where they are covered over by saline clay and silt eluvium. The bedrock contains heavy minerals and water-soluble salts which are sources of dust for dust storms. It is necessary to determine the classification so as to be able to identify sources of particulates in the atmosphere: sand >50 μm, silt 1-50 μm, clay <2 μm.

The wind-erodible fractions of 1-100 μm are as follows:

light or fine sand 50-100 μm
coarse silt 10-50 μm
fine silt 1-10 μm

In a normal atmosphere the largest amount of particles is in the form of a fine silt (1-5 μm). In a dusty atmosphere the largest amount of particulates is fine to coarse silt (5-50 μm). Mixtures of these airborne fractions generate variations in soils. The central wind-blown drifting sandy areas consist of a mixture of coarse and fine sands. The periphery of these sandy areas, especially the semi-fixed areas, consist of loamy sands (clay <20%, sand >50%) and sandy loam (clay 20-35%, silt <25%, sand >45%). They are a major source of dust during wind storms. On marginal plains loamy silt (clay 7-27%, silt 28-50%, sand < 52%), silty loam (silt >50%, clay <12%) and silty clay loam (clay 27-40%, sand <20%) are also picked up by the winds and put into the atmosphere. Besides these, more small clay particles (<5 μm) containing sulfates are also lifted by the winds [1, 6, 7].

Dynamics of Soils

The poor management of natural resources in the Black Lands during the 1970s and 1980s led to ecological catastrophe. In this period the productivity of pastures dropped by 3 to 5 times, the area of bare sands increased by 16-20 times, the areal extent of saline, waterlogged, and polluted soils also increased. The frequency of dust storms sharply increased, the albedo (reflectivity) of the land surface increased, the economy became depressed, and human health conditions worsened. The degradation of vegetation led to micro-climatic changes. During the hot summers, unvegetated areas (bare soils, low and sparse plants), when compared to vegetated areas (with trees, shrubs), increased the wind velocity by 2 to 3 m/s, the temperature of air by 5.5°C, the temperature of the soils by 12°C, and decreased the relative humidity by 13%. Sometimes the surface water, which has a high concentration of mineral salts, passes over the soils. During the warm period, soda saline soils are formed, and at the border of such patches a special kind of greenness occurs as a result of growing vegetation. The satellite images of these patches appear to have a higher reflectivity with identifiable boundaries.

The use of this characteristic to define the dynamics within the test site has been considered by Vinogradov et al. [2, 3, 4]. These patches were generated on pasturelands that had been heavily overgrazed by cattle, during the 1950s and 1960s. At the beginning of the 1950s these areas had provided good harvests, but, after 3 to 5 years, the soils became useless for agriculture. The soil at the surface contained considerable amounts of clay and minerals.

The sandy areas which are the sources of the largest amount of atmospheric pollution can be divided into three types. The first type is the old cumulative form with a spatial dimension of a few tens of hectares. A second type of sandy area is created by human activity, e.g., buildings and infrastructure. Its dimension is about 100 hectares. A third type is a natural sandy area, the dimensions of which can vary from 200 to about 1300 hectares. The changes of various types of soils within the test site during the last decade are shown in Fig. 1.

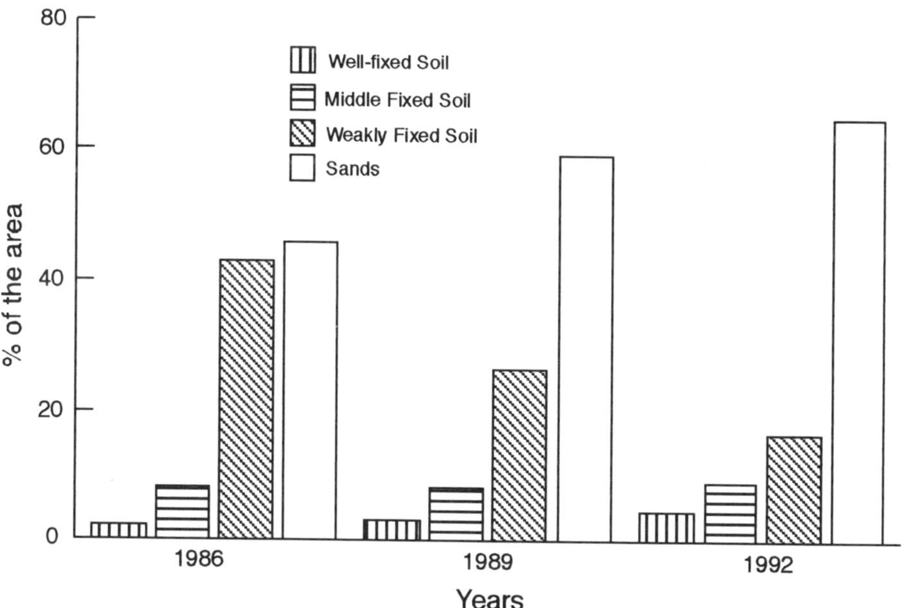

Fig. 1. Changes of various types of soil within the test site during last decade.

Preliminary Results

The July 1995 expedition was a preliminary one, undertaken for the purpose of choosing the test site and to test the equipment that had been installed in the AN-2 aircraft. The tests in the aircraft included the following:

- *aerosol samplers* - to define the mass concentration of aerosols and its size distribution;
- *photoelectric counters* - to define aerosol size distribution;
- *lidar* - to define the concentration and the optical features of aerosol clouds;
- *video camera* - to visualize the aerosol clouds and the location of the dust formation zone.

The characteristics of the measuring equipment have been discussed by Gorelik et al. [5]. Measurements were taken on 19, 20 and 22 July during three flights which were taken under different synoptic conditions. The flight of July 19 was carried out under weak gradient anticyclonic conditions. During the second flight (July 20), the weak gradient pressure field was displaced over a large area. A considerable wind shear was observed in the lower troposphere: the eastern wind with a velocity of 5 m/s at a height of 200 m changed to a northern one with the velocity of more than 10 m/s at a height of 2000 m. On July 21 a cold front with precipitation crossed Kalmykia. The third flight (July 22) was carried out after the rain; the weather was typical following such a front. These conditions are typical for Kalmykia.

During each flight, wind-blown dust and salt were observed, and in two flights it was observed in the same location. The aerosol concentration was measured at 0.15-0.25 mg/m^3. This value exceeded the background concentration that had been measured in the Aral region. The spatial distribution of the aerosol concentrations is shown in Figs. 2 and 3. Most of the aerosols were the particles of 0.5-0.1 μm. The relative amount of dust particles 0.5-5 μm in size was greater at the source of dust emissions. Lidar measurements identified the existence of the aerosol layers with relatively high concentration. The fluxes in aerosols are connected both with dust storms caused by the moving cold front and with thermal processes.

Temperature gradients can form a strong vortex structure that increases the "lift-off" of sand and its injection into the atmospheric layers above. We found qualitative support for this hypothesis during this field expedition. Our video showed the location of the dust plumes, identified the distinct phases and the convective "lift-off" along the boundaries of the zones with differing albedo (ground observations will be organized in these places). A detailed study of these processes, including ground observations and measurements of the physical and chemical properties of aerosols, were to be studied during the 1996 expedition.

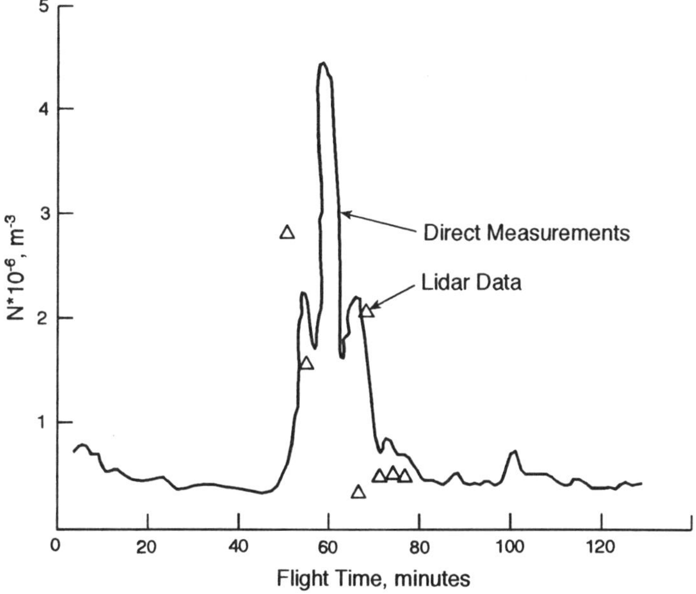

Fig. 2. Lidar and direct measurements of aerosol concentration during the flight.

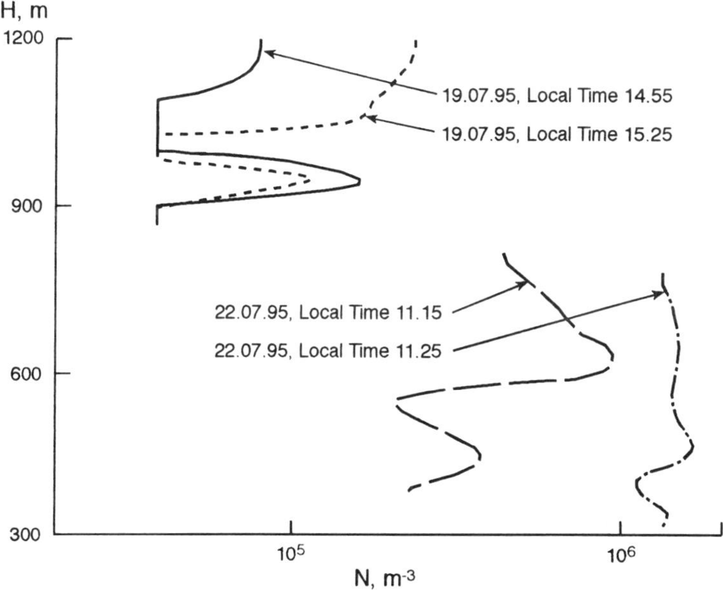

Fig. 3. Vertical profiles of aerosol concentration.

Conclusions

The Black Lands of Kalmykia are a major source of dust and salt in the atmosphere [8]. The aerosols include many salts that can affect agriculture and the local microclimate. The main source of dust particles is the newly eroded surface with high silt content and a destroyed vegetation cover. The methodology and equipment have been developed to measure parameters of dust emissions and soil dynamics. The dynamics show a tendency of the soil to erosion. To be able to forecast changing soil conditions would be of great value for land management.

Key Points

- The development of a monitoring system for the ultimate purpose to control processes related to the formation and transport of aerosols in semiarid areas and to reduce their influence on the environment and climate enables us to identify linkages between soil conditions, meteorological conditions and aerosol transport. The monitoring system requires ground measurements and space observations.

- The development of mathematical models will allow us to estimate the deposition of Kalmykian aerosols in Europe and its influence on the environment and climate.

- The dynamics of the Black Lands are affected by changes in groundwater that result from irrigation practices and the processes related to Caspian Sea level rise. The study of this influence is necessary for defining the strategy of land management.

References

1. Barenblatt, G.I., and G.S. Golitsyn, 1974: Local structure of natural dust storms. *Journal of Atmospheric Science*, **31**(10), p. 1917.

2. Vinogradov, B.V., V.V. Lebedev, K.N. Kulik, and A.N. Kaptzov, 1986: Measurement of ecological desertification from repeat aerospace photography. *Proceedings, USSR Academy of Sciences*, **285**, Biological Sciences Section, p. 690.

3. Vinogradov, B.V., D.E. Frolov, K.N. Kulik, 1991: Estimating instability in Black Earth ecosystems in Kalmykia from long series of aerospace measurements. *Proceedings, USSR Academy of Sciences*, **316**(1-6), Biological Sciences Section, 18-20.

4. Vinogradov, B.V., and V.V. Shakin, 1995: Logistic analysis for quantitative classification of indicators of areas of land degradation. *Doklady Biological Sciences*, Moscow, **341**(1-6), p. 404.

5. Gorelik, A.G., G.I. Gorchakov, I.G. Granberg, P.O. Shishkov, and P.V. Zakharova, 1996: Lidar-radiometric method for determination of the aerosol distributions with height in the lower atmosphere and experience using by example in aircraft expedition in Kalmykia 1995. *The Eighth International Symposium on Acoustic Remote Sensing and Associated Techniques of the Atmosphere and Oceans.* Held in Moscow, Russia in May 1996. Moscow, Russia: Obukhov Institute of Atmospheric Physics, Russian Academy of Sciences, and Russian Acoustic expedition.

6. Gillette, D.A., 1986: *Wind Erosion in Soil Conservation.* Washington, DC: National Academy Press.

7. Gillette, D.A., 1988: Threshold friction velocities for dust production for agricultural soils. *Journal of Geophysical Research,* **93**(D10), p. 12645.

8. Granberg, I., V. Ponomarev, A. Eidelman, and M. Nagorny, 1996: Modeling of admixture transport in atmosphere and MHD-flow. *The Eighth Beer-Sheva International Seminar on MHD-Flows and Turbulence.* Held in Jerusalem, Israel February 1996. Jerusalem: Israel Academy of Sciences and Humanities, p. 173.

IV: Coping with Environmental Change

SOCIO-ECOLOGICAL CHARACTERIZATION: A FRAMEWORK FOR ECOSYSTEM MANAGEMENT INFORMATION

JOHN J. KINEMAN
National Geophysical Data Center
National Oceanic and Atmospheric Administration
Boulder, Colorado, USA

BRADLEY O. PARKS
Cooperative Institute for Research in the Environmental Sciences
University of Colorado
Boulder, Colorado, USA

Introduction

Practical methods are needed to communicate information about the health and sustainability of ecosystems that are under increasing human pressure, and to evaluate choices from the combined perspectives of science, society, and management. Ecological characterization is an established approach that may be enhanced to meet these evolving needs. This chapter discusses the concept and content of ecological characterization, focusing selectively on improvements in the production and application of its resulting information products.

Methods and Practice

The term "ecological characterization" was adopted in the 1970s by the US Fish and Wildlife Service's Coastal Ecosystems Group (and collaborators) to describe their rigorous efforts to integrate information useful for administering marine oil and gas leases in the Outer Continental Shelf (OCS) and in planning for associated nearshore and onshore impacts [3]. The practice of ecological characterization met increasing need for environmental and ecological information synthesis to support management and planning activities. Although its origins were in marine minerals management with contributions from other fields, the approach may be generalized to other systems and issues, including the health and sustainability of ecosystems. Ecological characterizations of coastal ecosystems have been conducted as intensive descriptions of many kinds of ecosystems and management units, and their results have been published for limited distribution as printed documents.

M.H. Glantz and I.S. Zonn (eds.),
Scientific, Environmental, and Political Issues in the Circum-Caspian Region, 157-164.
© 1997 *U.S. Government. Printed in the Netherlands.*

In a recent workshop on ecological characterization [5], a review panel emphasized that greater emphasis should be given to societal elements, both as an object of analysis and as an active element in a larger valuation and decision-making process. Because of this emphasis, the term "socio-ecological characterization" has been suggested either as an expansion of the historical approaches to ecological characterization, or at least as a reminder of the importance of societal factors in any such effort. In either case, dealing with the combined influences of ecology, environment, management, human activity, and societal values is strongly implied if not explicitly represented. By bridging disciplines, and linking their data, information, and models, ecological characterization can deal with a large range of societal issues. The workshop also emphasized that ecological characterizations should address ecosystem-based management, to improve understanding of ecological and socio-economic sustainability, explicitly considering present and future generations. This means balancing evolving natural factors and human values in a stable manner — a challenging task that will require better knowledge and understanding, informed valuation, and suitable management tools.

Applications — Past and Present

Ecological characterization literature includes many useful examples treating different systems and management units from rivers, bays, and estuaries to watersheds and landscapes. Some modern examples include participation by individuals and local constituencies that have a "stake" in the outcome of management efforts [6]. Each uses slightly different methods and techniques but all share an emphasis on understanding systems rather than their separate parts and on decision making that is informed by the best possible synthesis. Although they have sometimes used computerized systems for data preparation and analysis, early ecological characterizations were (of necessity) published and delivered as paper documents [4, 1, 10], which represented the combined synthesis efforts of many scientists and managers. In paper form, they were unavoidably static, one-time documents that were difficult to update, reproduce, or apply to new questions. [5, 7].

With recent availability of digital publication techniques (on CD-ROM and the World Wide Web), it is now possible to provide ecological characterization information to users in an interactive form that is far more flexible, adaptable, and maintainable than former documents. Data, prepared information, and highly interactive analytical tools can all be provided in one product that in turn can support flexible information retrieval, data exploration, analysis and various forms of modeling. This allows the user to reevaluate the data and information base, or ask different questions. As a more flexible resource, ecological characterization products can have potentially wide use and long-term value.

Based on the workshop recommendations and programmatic needs of a newly formed Coastal Services Center within the U.S. National Oceanic and

Atmospheric Administration (NOAA), a digital prototype, the "Ecological Characterization of Otter Island, South Carolina" [2] was produced and used to test new approaches. This product demonstrated the means and utility of publishing ecosystem synthesis information in digital form, including digital text, images, bibliographies, tabular and geographic data, management protocols and scenarios, and other relevant information. It also provided a limited basis for evaluating the socio-ecological characterization approach described above. The intended benefits were to simplify access and updating of data and information, broaden distribution, foster wider use, and stimulate improvements at many levels. Based on the prototype experience, additional characterizations are planned for coastal sites around the U.S. Similar approaches may be useful elsewhere and in different ecosystems.

Content and Scope

Ecological characterization, and its "socio-ecological" counterpart, are presented here as structured means for assembling and presenting current knowledge about ecosystems and human activity for management purposes. Although implementations vary, ecological characterizations synthesize interdisciplinary knowledge about ecosystems including their biological resources, physical environment, key processes and important human interactions (This is distinct from another common use of the term "characterization," which refers to a single set of environmental characteristics or theme, such as "land cover characterization," distributed as data sets for more general use). Ecological characterizations may include descriptions or analyses of anticipated resource uses, proposed development scenarios, and management response options. But the main objective of the approach described here is to involve users with the raw material (data and information) for understanding ecosystem management and to give them the means to derive and compare alternative outcomes.

One result of this work was to confirm three traditionally prescriptive elements of an ecological characterization, and to suggest a fourth that is explicitly emphasized by socio-ecological characterization.

I. It should provide a clear management orientation identifying the issues being considered and the overall goal of management options, incorporating the perspectives of societal impacts and valuation. This may include the presentation of management plans and future scenarios.

II. It should present a coherent picture of the ecosystem and human involvement, including the system's constituent parts and functional ecology. This may include the use of models.

III. It should reference or incorporate a comprehensive data and information base. This should include equally comprehensive metadata and documentation, along with key references, bibliographies, and various data access tools, as needed.

Given the opportunity to provide integrated databases in digital form, it is now possible to give users direct capabilities to manipulate the database for analysis, modeling, validation, and other purposes.

IV. It should represent and support a larger and more societally based process of resource valuation, community planning, and reconciliation of competing interests among "stakeholders."

Since ecosystem information can be extremely complex even for simple questions, it is important to limit an ecological characterization to those elements that address key system factors. The prescriptive elements listed above establish this focus. First, characterization information is focused by the management goals, questions, and issues of primary interest. Second, a characterization is necessarily limited by present understanding of ecosystem function. This may be represented by a generalized concept of the important components and interactions (including human factors); sometimes expressed as a "conceptual model." Third, a characterization is based on current knowledge; it is not a program for continued study. As such, it is limited to synthesis of readily available data and information, human use and management scenarios, and description or models of processes, prioritized by the first two elements above. Ecological characterization (or its expanded form) may require field work to supplement or validate observations and may identify data and information gaps, but should be aimed at practical application in the immediate future. Finally, it is prepared within the context of its intended use, to function primarily as a device for active communication between various interests.

Challenges

Several important challenges remain regarding creation of these contemporary characterizations as interactive digital publications. The most obvious is how to organize the data, information and tools for effective use. Another is how to ensure appropriate quality and reliability of the data and information for multiple purposes, especially when combining information (synthesis) to help understand the behavior or condition of a given system, or to evaluate management options. There is also the question of how to deliver this information in a form that can be used routinely by practitioners and yet one that preserves the full potential that digital processing capabilities offer to more demanding users who may wish to extend the analyses or explore new questions.

ORGANIZATION

A good way to organize ecosystem information, especially with management goals in mind, is with a spatial (or geographical) perspective. Geographic Information Systems (GISs) provide spatial organization of data which allows the description of phenomena within the context of their surroundings. They also aid in the analysis of spatial relationships and processes. If used, GISs must be linked selectively to other relevant kinds of information such as non-spatial (non-map) information or predictions derived from mathematical and simulation models [8, 9]. GISs can provide much-needed capabilities to the information user ranging from data retrieval operations to analytical procedures letting the user ask "what-if" questions regarding management options or scenarios for future change. Even if users do not exploit their more technical capabilities, GIS tools can provide guided explorations of the ecological characterization information base through tutorial-like presentations. The great value of GISs is that they provide a common structure for quantifying spatial objects and their relationships which can be shared across disciplines.

Quality and Usefulness

The problem of delivering quality information for flexible use must be addressed by providing adequate ancillary information about the ecological characterization information base. "Data about data," or *metadata,* incorporate information about the form, content, quality, and condition of data and included information. Less structured *documentation* can provide needed insights about the design, use, and application of the data and information, or about assumptions or limitations. Proper use of ecological characterization information for management and decision making requires that the user consider the limitations built into databases during data collection and preparation, such as collection techniques, timing and frequency of collection, resolution, sample locations, and other design factors. A variety of ancillary information can be provided as part of metadata and documentation, including system-defined metadata files and lineage records, descriptive texts, bibliographic information, and key literature reprints. One should also be aware that use of model-derived data, information, and scenarios to represent new indices or change in a system through time requires information about the starting assumptions and initial conditions used by the models.

User Interface and Software Tools

The challenge of delivering useful products for users with a range of different needs and skills means balancing the need for simplicity and ease of use with the need for higher levels of functionality. To accomplish this, a modular user interface can be created which links to various sub-components, including calls to independent software tools. For example, a relatively friendly and Web-compatible interface can be created using an HTML (HyperText Markup Language) browser. This kind of interface has the ability to launch hypertext as well as other software to implement various capabilities, such as a bibliographic data browser, a Geographic Information System, video or still picture presentations (including audio), and document viewers. Various levels of GIS capability can be included, from simple browse and retrieval to interactive analysis and static spatial modeling. Finally, export capabilities can provide valuable links to other software systems using standard exchange formats. The fact that it is possible to provide capabilities at many levels in a single product, without compromising data content or quality, is clearly an advantage of digital publication.

Conclusion

Socio-ecological characterization provides a structured approach to the synthesis of societal and ecological information for management purposes. Digital publication provides new options for organizing and presenting such information so that it can be applied to changing ecosystem management needs. Combining natural science, resource, human use, and management information in an interactive format with analytical capabilities enhances the user's capability to compare information across themes, through time, between locations, and at different scales. In addition to being applicable to different questions, this approach offers new possibilities for improving information content, since the means of production promises products that can be updated, revised, and reissued. This flexibility supports adaptive management goals by providing tools for planning and for retrospective analysis, and the means to incorporate results from such exercises into future product releases.

We must recognize, however, that the use of the ecological (or socio-ecological) characterization approach implies an inclusive interaction that involves scientific, resource management, educational, and all other users or "stakeholders," with appropriate feedbacks between them. While this has not been elaborated here, we have described tangible improvements to the delivery of products that may support a larger process. Existing practices that deal with values and analysis of management results, such as *adaptive management* and similar approaches, may also be part of this larger process.

More work is clearly needed to refine the contents, methods, information capture techniques, and delivery mechanisms for ecological characterization. An important limitation of this practice will continue to be the ready availability of data and information for synthesis. Nonetheless, we believe that such work can and should be actively pursued, and that a comparative approach should be used to test alternative methods in different ecological circumstances.

References

1. Barclay, Lee A., 1978: An Ecological characterization of the sea islands and coastal plain of South Carolina and Georgia. In: Coastal Zone '78: Symposium on Technical, Environmental, Socioeconomic, and Regulatory Aspects of Coastal Zone Management. Proceedings of a meeting held in San Francisco, CA. American Society of Civil Engineers, New York. pp 711-720.

2. South Carolina Department of Natural Resources (SCDNR), USDOC/NOAA Coastal Services Center, and USDOC/NOAA National Geophysical Data Center, 1996: Ecological characterization of Otter Island, South Carolina: A prototype for interactive access to coastal management information. US Department of Commerce (USDOC), National Oceanic and Atmospheric Administration (NOAA), Coastal Services Center (CSC), Charleston, SC. NOAA CSC /7-96/001. Digital information on CD-ROM. 236 megabytes.

3. Johnston, James B., 1982: Managing coastal ecosystems: progress towards a systems approach. IN: Research on fish and wildlife habitat. EPA 600/8-82-022. United States Environmental Protection Agency. Washington, DC. pp. 47-57.

4. Johnston, James B., 1978: Ecological Characterization - an overview. IN: Coastal Zone '78: Symposium on Technical, Environmental, Socioeconomic, and Regulatory Aspects of Coastal Zone Management. Proceedings of a meeting held in San Francisco, CA. American Society of Civil Engineers, New York. pp. 692-710.

5. Kineman, John J. and Bradley O. Parks, 1996: Ecological characterization: Recommendations of a NOAA Science Advisory Panel. Workshop held March 9-10, 1995 at the National Center for Atmospheric Research (NCAR), Boulder, CO. Sponsored by the NOAA Coastal Services Center (CSC) and convened by the NOAA National Geophysical Data Center (NGDC) and University of Colorado Cooperative Institute for Research in the Environmental Sciences (CIRES). US Department of Commerce, National Oceanic and Atmospheric Administration, National Geophysical Data Center, Boulder, CO and Coastal Services Center, Charleston, SC. NOAA NGDC KGRD#32, NOAA CSC/10-96/001. 48p.

6. Marshall, William D., Ed., 1993: Assessing change in the Edisto River Basin: An Ecological Characterization. Report No. 177. South Carolina Water Resources Commission. Columbia, SC. 149 pp.

7. Parks, Bradley O., and John J. Kineman, 1996: Implementing a Coastal Ecological Characterization Prototype: Proceedings of an Analysis and Characterization planning workshop. Workshop held March 15-17, 1995 at Seabrook Island, SC. Sponsored and convened by the NOAA Coastal Services Center (CSC), in collaboration with the NOAA National Geophysical Data Center (NGDC), and the University of Colorado Cooperative Institute for Research in the Environmental Sciences (CIRES). NOAA CSC/10-96/002,. Charleston, SC. 52p.

8. Parks, Bradley O., 1993: The need for integration. In: Goodchild, M.G., Parks, B.O., and Steyaert, L.T. (Eds) Environmental Modeling with GIS. Cambridge: Oxford University Press. pp 31-34.

9. Susan G. Stafford, James Brunt, and William Michener, 1994: Integration of scientific information management and environmental research. In: Michener, W.K., Brunt, J.W., and Stafford, S.G. (Eds) Environmental information management and analysis: ecosystem to global scales. London: Taylor & Francis, pp 3-19.

10. Watson, Jay F., 1978: Ecological Characterization of the coastal ecosystems of the United States and its territories. In: Lindstedt-Siva, J. (Ed) Energy/Environment '78. Proceedings of a symposium on energy development impacts. Society of Petroleum Industry Biologists. (Los Angeles, CA) pp 47-53.

AN OVERVIEW OF GLOBAL ENVIRONMENTAL CHANGE RESEARCH

JOÃO M.F. MORAIS
International Geosphere-Biosphere Programme
IGBP Secretariat
The Royal Swedish Academy of Sciences
Stockholm, Sweden

Since February 1996, the International Council of Scientific Unions (ICSU) has been a co-sponsor of the International Human Dimensions Programme on Global Environmental Change (IHDP), in partnership with the International Social Science Council (ISSC). Thus, ICSU sponsors all three international global environmental change programs: the International Geosphere-Biosphere Programme (IGBP), the World Climate Research Programme, and now the IHDP.

Introduction

Over the past few years, the third IHDP Symposium (Geneva, August 1995), the IGBP's Fourth Scientific Advisory Council (Beijing, October 1995) and First Congress (Bad Münstereifel, April 1996), and the workshop to launch an International Research Institute for climate prediction (Washington DC, November 1995), each made clear the crucial role of social science research in international research programs focused on the study of the physical and biogeochemical aspects of the Earth system.

This awareness draws from the fact that humans are bound to their physical and biological environments in terms of shelter, territory, food and water. At the same time, human practices of resource utilization have impacts on natural processes, with significant consequences for the carrying capacity of the environment, and its long-term sustainability. The term *Global Change* in an anthropocentric context was first used in political science in the 1970s, anticipating such concepts as "sustainable development" and "Earth System Science." However, the meaning of "Global change" was expanded from the early 1980s to a meaning related to geocentric use which has since become dominant [25]. It is, therefore, timely to develop integrative approaches (biogeochemical, physical *and* human) [18, 26], where the role of social science is not marginal to the process (see the attached Fig. 1 "Bretherton" diagram, adapted from Williamson [29, 1]).

M.H. Glantz and I.S. Zonn (eds.),
Scientific, Environmental, and Political Issues in the Circum-Caspian Region, 165-183.
© 1997 *Kluwer Academic Publishers. Printed in the Netherlands.*

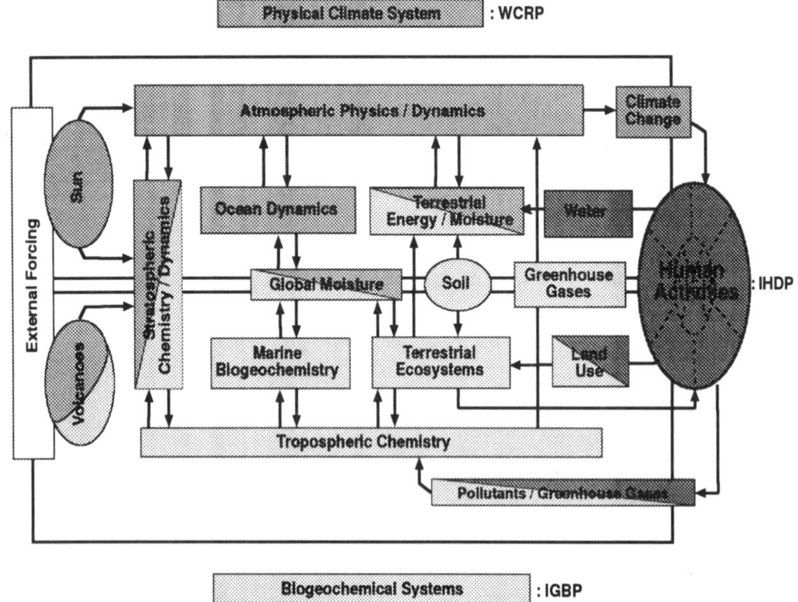

Fig. 1. A modified "Bretherton diagram" to better highlight human activities

The fundamental nature of the interactions between humans and the Earth system is not different from the past. However, the spatial and temporal scales of current impacts are unprecedented. Humans have been engaged since the dawn of time in finding creative solutions to such problems. This process has cut across biological and cultural differences, as well as methodological ones. It has been on "trial and error" in which a few new responses slowly arise from a considerable number of "experiments." Science offers a better chance of success, provided that it builds upon *systemic* and *diversified* networks which link the natural and social sciences, and addresses comprehensively both basic and applied approaches, and both quantitative and qualitative analyses. In general, this will require a shift of existing attitudes to ensure that knowledge is articulated between the science community, governmental and non-governmental institutions and the public, creatively participating in determining better policies and focused choices. These will derive from a more effective use of information, improved educational resources, and well focused research and development. Additionally, the science itself will benefit from the plurality of knowledge emerging from within a variety of different and original viewpoints, thereby avoiding the dangers associated with pursuing a monolithic approach. The ultimate goal of science is not science itself, but a better understanding of a world which has no divisions.

It is crucial that global change processes are addressed globally. Local, regional and international organizations have to interact and collaborate fully, and research projects have to be designed to address pressing solutions to socio-economic and cultural problems that are occurring worldwide. Public opinion increasingly requires that science should be useful, and its benefits more tangible. Communities are pressing for improved living standards and security. Policy making calls for rapid solutions, requiring a quick return on the investment in science research. However, scientists cannot promise the desired results either to a schedule or even with guaranteed certainty. There is a growing acceptance among scientists of the need to make scientific results clearer to nonspecialists and to provide inputs to the policy process.

Global Change and the IGBP

The International Geosphere-Biosphere Programme (IGBP) was established in 1986 by the 21st General Assembly of the International Council of Scientific Unions (ICSU). It was conceived as a sister program to the World Climate Research Programme (WCRP) and the International Human Dimensions of Global Environmental Change Programme (IHDP) (see Fig. 1). The IGBP's goal is "*to describe and understand the interactive physical, chemical and biological processes that regulate the total Earth system, the unique environment it provides for life, the changes that are occurring in this system, and the manner in which they are influenced by human activities.*" Priority in the Programme is placed on "those areas of each of the fields involved that deal with key interactions and significant changes on time scales of decades to centuries, which most effect the biosphere, which are most susceptible to human perturbation, and which will most likely lead to practical, predictive capability."

The Programme is organized in a number of Core Projects (CP), each addressing a scientific question of major importance for the understanding of the role of biogeochemical cycles in the field of global change research. Eight CPs have been established to date, and an additional one is in the planning stage. In addition, three overarching Framework Activities (FA) have been established with the objective of achieving a high degree of integration and synthesis of results at the Programme level, and to ensure maximum participation by scientists worldwide. Increasingly, CPs and FAs are forming joint activities to address problems which crosscut their interests. For example, Biospheric Aspects of the Hydrological Cycle Core Project (BAHC), International Global Atmospheric Chemistry Core Project (IGAC), and Global Change Terrestrial Ecosystems Core Project (GCTE) developed a plan for a series of terrestrial transect experiments. This is now being developed into an IGBP-wide terrestrial transacts plan.

National committees of participating countries constitute the members of the IGBP. These represent and coordinate global change research activities in their respective countries. At present there are 73 National committees [16].

In brief, the IGBP Core Projects and Framework Activities are as follows [11]:

(I) BIOSPHERIC ASPECTS OF THE HYDROLOGICAL CYCLE CORE PROJECT (BAHC)

Primary science question: How does vegetation interact with physical processes of the hydrosphere?

Objectives:

- to determine the biospheric controls of the hydrological cycle through field measurements, for the purpose of developing models of energy and water fluxes in the soil-vegetation-atmosphere system at spatial scales ranging from vegetation patches to GCM grid cells and at various temporal scales;

- to develop appropriate data bases that can be used to describe the interactions between the biosphere and the physical Earth system, and to test and validate model simulations of such interactions.

Organization: There are four foci addressing the development, testing and validation of one-dimensional soil-vegetation-atmosphere transfer (SVAT) models, regional-scale studies of land surface properties and fluxes, temporal and spatial diversity of biosphere-hydrosphere interactions, and the downscaling of outputs from General Circulation Models to regional scales (the "weather generator" project) [10].

Status: the project is well advanced in its implementation phase. BAHC activities are being closely coordinated with those of the World Climate Research Programme (WCRP) core project Global Energy and Water Exchange (GEWEX). Results range from advances in the understanding of how to aggregate land-surface properties using simple averaging rules to detailed new knowledge about energy and water fluxes in semiarid areas (based on major field experiments in the Sahel and in Spain). A major scientific symposium was held in Hamburg in mid-1995 to present BAHC results.

(II) GLOBAL CHANGE TERRESTRIAL ECOSYSTEMS CORE PROJECT (GCTE)

Primary science question: How will global changes affect terrestrial ecosystems, and what will be the feedbacks to the climate system?

Objectives:

- to predict the effects of changes in climate, atmospheric composition, and land use on terrestrial ecosystems, including agricultural and forest production;

- to determine how these effects lead to feedbacks to the atmosphere and physical climate system.

Organization: There are four foci addressing ecosystem physiology, change in ecosystem structure, global change impact on agriculture and forestry, and global change and ecological complexity [9].

Status: the project is well advanced in its implementation phase, except for Focus 4 (on ecological complexity), which is still at an early stage of development. Results range from new knowledge of the effects of enhanced CO_2 levels on plant growth and carbon sequestration, to improved understanding of terrestrial ecosystems as sources and sinks of atmospheric CO_2. A major scientific symposium was held in Woods Hole, Massachusetts, USA in late 1994 to present GCTE results.

(III) INTERNATIONAL GLOBAL ATMOSPHERIC CHEMISTRY CORE PROJECT (IGAC): jointly sponsored with the Commission on Atmospheric Chemistry and Global Pollution of the International Association of Meteorology and atmospheric Sciences.

Primary science question: How is the chemistry of the global atmosphere regulated, and what is the role of biological processes in producing and consuming trace gases?

Objectives:

- to develop a fundamental understanding of the processes that determine the chemical composition of the atmosphere;
- to understand the interactions between atmospheric chemical composition and physical, biospheric and climatic processes;
- to predict the impact of natural and anthropogenic forcing on the chemical composition of the atmosphere.

Organization: There are eight foci of which five are regional (addressing marine, tropical, polar, boreal and mid-latitude regions), two are general (global trends, distributions, transports, transformations, sinks; fundamental, covering measurement calibration and intercomparisons, laboratory studies and new instrument development), and one covers the study of tropospheric aerosols [13].

Status: the project is well advanced in its implementation phase. IGAC recently merged with the International Global Aerosol Programme (IGAP), adding substantially to its strength in tropospheric aerosol research. The World Climate Research Programme (WCRP) has recently requested to become a cosponsor of the project. Results include new information on trace gas emissions from biomass burning to studies of the mitigation of methane emissions from rice paddies. Two major scientific symposia have been held to present IGAC results, one in Eilat, Israel in 1992, and one in Fuji-Yoshada, Japan in 1994.

(IV) LAND OCEAN INTERACTIONS IN THE COASTAL ZONE CORE PROJECT (LOICZ)

Primary science question: How will changes in land use, sea level and climate alter coastal ecosystems, and what are the wider consequences?

Objectives:

- to determine at regional and global scales the fluxes of material between land, sea and atmosphere through the coastal zone, the capacity of coastal ecosystems to transform and store particulate and dissolved matter, and the effects of changes in external forcing conditions on the structure and functioning of coastal ecosystems;

- to determine how changes in land, sea, climate, sea level and human activities alter the fluxes and retention of particulate matter in the coastal zone, and affect coastal morphodynamics;

- to determine how changes in coastal systems, including responses to varying terrestrial and oceanic inputs of organic matter and nutrients, will affect the global carbon cycle and the trace gas composition of the atmosphere;

- to assess how the responses of coastal systems to global change will affect the habitation and usage by humans, of coastal environments, and to develop further the scientific and socio-economic basis for the integrated management of the coastal environment.

Organization: There are four foci addressing the effects of changes in external forcing or boundary conditions on coastal fluxes, coastal biogeomorphology and global change, carbon fluxes and trace gas emissions, and the economic and social impacts of global change in coastal systems [14].

Status: The project recently entered its implementation phase. A major open meeting held in Quezon City, the Philippines, attracted over two hundred international scientists to discuss the LOICZ science agenda.

(V) LAND-USE AND LAND-COVER CHANGE CORE PROJECT (LUCC): jointly sponsored with the International Human Dimensions Programme (IHDP) on Global Environmental Change.

Primary science question: In what ways are land cover and land-use practices evolving, and how are such changes affecting climate, biogeochemistry, and the composition and structure of ecosystems?

Objectives:

- to determine how land cover has changed over the past 300 years;
- to determine the major human causes of land-use change in different geographical and historical contexts;
- to determine how changes in land use will affect land cover in the next 50-100 years;
- to determine how immediate human and biophysical dynamics affect the sustainability of specific types of land use;
- to determine how changes in climate and global biogeochemistry might affect both land use and land cover, and *vice-versa*.

Organization: There are three foci addressing land-use dynamics, land-cover dynamics, and regional and global models. In addition, two integrating activities addressing data and classification and scalar dynamics [15].

Status: The project was approved in late 1994 as a joint IGBP Core project and in early 1995 as an IHDP research program. Work has commenced on the preparation of an implementation plan.

(VI) JOINT GLOBAL OCEAN FLUX STUDY CORE PROJECT (JGOFS): jointly sponsored with ICSU's Scientific Committee on Ocean Research.

Primary science question: How do ocean biogeochemical processes influence and respond to climate change?

Objectives:

- to determine and understand on a global scale the processes controlling the time-varying fluxes of carbon and associated biogenic elements in the ocean, and to evaluate the related exchanges with the atmosphere, sea floor and continental boundaries;
- to develop a capacity to predict on a global scale the response of oceanic biogeochemical processes to anthropogenic perturbations, in particular, those related to climate change.

Organization: the project involves an extensive set of shipborne observations and process studies, remote sensing activities and model development.

Status: The project is well advanced in its implementation phase with a large number of internationally coordinated shipborne cruises having been completed and data sets produced, often in close collaboration with the WCRP core projects [the World Ocean Circulation Experiment (WOCE) and Tropical Ocean Global Atmosphere (TOGA; completed in 1994)]. Results include unique information on the fertilization of the upper ocean with iron, the classification of the world ocean into biological "provinces," and a new understanding of the export of dissolved organic carbon from the upper ocean. A major science conference was held in Villefranche-sur-mer, France in mid-1995 to present JGOFS results.

(VII) PAST GLOBAL CHANGES CORE PROJECT (PAGES)

Primary science question: What significant climatic changes and environmental changes occurred in the past, and what were their causes?

Objectives:

- to reconstruct the detailed history of climatic and environmental change for the entire globe for the period since 2000 BP, with a temporal resolution that is at least decadal and ideally annual to seasonal;

- to reconstruct a history of climatic and environmental change through a full glacial cycle, in order to improve our understanding of the natural processes that invoke global climatic changes.

Organization: There are three research foci addressing Global Palaeoclimate and Environmental Variability, Palaeoclimate and Environmental Variability in the Polar Regions, and Human Interactions in Past Environmental Changes respectively; a Climate System sensitivity and Modeling Focus; and a cross-project Focus addressing analytical and Interpretative Activities. Within each of the observational foci, the research is split into two "temporal streams." Temporal stream I focuses on the reconstruction of detailed history of climatic and environmental change for the entire globe for the period since 2000 BP, with temporal resolution which is at least decadal and, ideally, is annual or seasonal. Temporal Stream II focuses on glacial-interglacial cycles over the last several hundred thousand years with at least a century resolution. The objective is to understand the dynamics that cause large-scale natural variations [8].

Status: The project is well advanced in its implementation phase. Results include the discovery of climate "flickering" in the Eemian (from ice core data) and apparent rapid changes in atmospheric CO_2 and CH_4 content linked with changes in the intensity of the deep ocean "conveyor belt" (from peat bog and marine sediment data).

(VIII) FUTURE CORE PROJECTS

Two additional marine projects are under development. The first, The Global Ecosystem project (GLOBEC) addresses the ocean food web and the way this will be affected by, and may affect, global change. The GLOBEC science plan was approved by the IGBP Scientific Committee at its meeting in Beijing in November 1995. The second project, previously known as the Global Ocean Euphotic Zone Study (GOEZS) but now renamed the Surface Ocean Lower Atmosphere Study (SOLAS), is concerned with change and feedbacks associated with the coupled physics, chemistry and biology of the upper ocean, and is developing its science plan.

(IX) DATA AND INFORMATION SYSTEM FRAMEWORK ACTIVITY (DIS)

Goal:

To improve the supply, management and use of data and information that are needed to attain IGBP's goals.

Objectives:

- to carry out activities directly leading to the generation of relevant data sets and other data system components for IGBP;
- to ensure the development of effective data and information management systems for IGBP, including the preparation and implementation of a program-wide Data System Plan, and assisting IGBP Core Projects in the development of their individual Data System Plans;
- to assist in meeting the data and information needs of IGBP through partnerships and collaborations with other organizations and agencies.

Organization: There are three foci addressing Data set development, Data Management and Dissemination, and Data Co-ordination in an International Context [11].

Status: The activity is well advanced in its implementation phase. Results include the definition and implementation (through various agencies) of a global 1 km resolution data set of the Earth's terrestrial surface from the satellite instruments AVHRR, the definition and initial implementation of a global soil data set, the development and initial implementation of a global 1 km resolution land-cover classification, and the development and prototyping of a global fire data set derived from satellite data.

(X) GLOBAL ANALYSIS INTERPRETATION AND MODELING FRAMEWORK ACTIVITY (GAIM)

Goal:

To advance the study of the coupled dynamics of the Earth system using as tools both data and models.

Objectives:

- to implement a strategy for the rapid development, evaluation, and application of comprehensive prognostic models of the global biogeochemical system which could eventually be linked with models of the physical climate subsystem;

- to propose, promote, and facilitate experiments with existing models or by linking subcomponent models, especially those associated with IGBP Core Projects and with WCRP efforts. Such experiments would be focused on resolving interface issues and questions associated with developing an understanding of the prognostic behavior of key processes:

- to clarify scientific issues facing the development of global biogeochemical models and the coupling of these models to general circulation models;

- to contribute to international assessment exercises, particularly the Intergovernmental Panel on Climate Change (IPCC) process, by conducting timely studies that focus on elucidating important, unresolved scientific issues associated with the changing biogeochemical cycles of the planet and focus on the role of the biosphere in the physical climate subsystem, including its role in the global hydrological cycle;

- to advise the IGBP Scientific committee on progress in developing comprehensive global biogeochemical models, and to maintain scientific liaison with the WCRP Steering Group on Global Climate Modeling.

Organization: In its initial phase, GAIM is focusing on modeling. The effort is structured by topic and by time period. The topics are CO_2, trace gases, and climate-vegetation interactions. The time periods are paleo (<20 Ky..), fossil fuel (<200 y), contemporary (<20 y), and future [11].

Status: The activity is in its implementation phase. Six projects are currently under way:

- 6000 BP climate-vegetation interactions;
- regional interactions of climate and ecosystems;

- coupled atmosphere-land-ocean carbon system;
- changes in terrestrial carbon storage;
- atmospheric-ecosystem interactions on methane;
- ocean carbon intercomparison project.

A major science conference was held in Garmisch-Partenkirchen (Germany) in September 1995 to present early results.

(XI) SYSTEM FOR ANALYSIS, RESEARCH AND TRAINING (START): jointly sponsored by the WCRP and the IHDP.

Goal:

To promote interdisciplinary studies of relevance to global change science on a regional basis.

Organization: START has established a number of regional divisions in which it is developing research networks. It has an active program of capacity building, including workshops and summer schools, and is in the process of establishing a fellowships scheme [7].

Global Change and the International Human Dimensions Programme (IHDP)

The International Social Science Council (ISSC) created an *ad hoc* committee in 1986 to explore the possibility of developing an international social science research program that would parallel and complement IGBP's. At its First Scientific Symposium (Mallorca, 1990) the program was discussed and formally initiated [2] after identification of seven research topics. (See Fig. 2).

The HDP's Work Plan 1994-1995 [6] was further refined during the Second Scientific Symposium (Paris, December 1992) [4, 19], at the HDP's and ISSC's Officers, and Executive Committee meetings, respectively, in Mexico (Ensenada, January 1993) and Spain (Barcelona, May 1993).

The purpose and rationale of the HDP program was to foster "*activities that seek to describe and understand the human role in causing global environmental change and the consequences of these changes for society*," complementing IGBP's and WCRP's efforts, and overlapping with, but without duplicating, national programs [6].

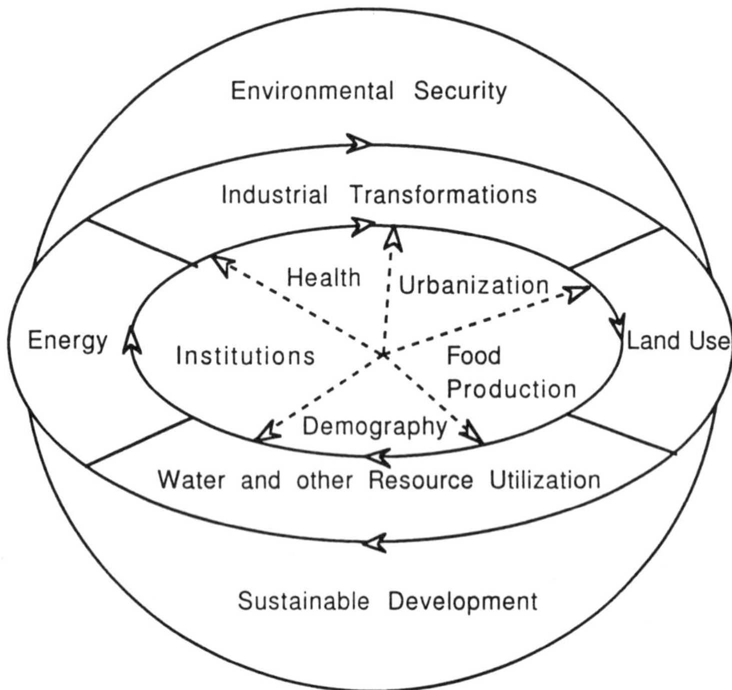

Fig. 2. A tentative schematic representation of socio-environmental research. Some examples of over-arching concerns, core research and cross-cutting themes are in concentric order.

The first efforts were addressed at evaluating data collection related to survey and public opinion research, population and economics [3,4,5]. The rationale for an international and interdisciplinary research effort was based on a number of assumptions:

- the need for links with the other sister programs on global environmental change (e.g., IGBP and WCRP);
- an emphasis on processes with global scope requiring data sets and interdisciplinary cooperation that exceed national capacities;
- a focus on problems that affect human survival and well-being, and for which national perspectives may produce an insufficient level of knowledge;
- a focus on feasible questions for which practicable research efforts can produce results over approximately the next decade;

- goals to advance the substance and the methods of social science through collection of global data sets, elaboration of new research questions, and exchange of methodologies among disciplines.

The HDP framework program [6] listed the following research areas:

- Land-use and land-cover change (LUCC), co-sponsored by both IHDP and IGBP.

- Industrial transformation and energy production and consumption.

- Demographic and social dimensions of resource use.

- Public attitudes, perceptions, behavior and knowledge, with sub-groups on "The Global Environmental Survey - GOES" and "Perception and Assessment of Global Environmental Change - PAGEC".

- Institutions.

- Environmental security and sustainable development.

The above topics were suggested by a number of research questions:

- What are the human driving forces of global change?

- How are these social processes linked to the physical processes of global change?

- What are the impacts of physical global change processes on human systems?

- What are the differential sensitivities of various social, political, technological and economic arrangements to global change processes?

- What is the potential for adaptation, and what mitigation strategies could different social systems formulate and adopt?

During the HDP Third Symposium (Geneva, September 1995), additional crosscutting themes were identified as of possible interest to the HDP program, such as Global Environmental Change and Environmental Security, Human Health, Vulnerability and Security, the Globalization of Trade, and Ethics. In order to revisit the progress being made, scoping reports were recently produced by the different working groups [20, 21, 22, 23, 27, 28, 30, 31, 32].

As a result of the recent reorganization of the IHDP (which since February 1996 has been cosponsored by both ICSU and ISSC), a new Scientific Committee was appointed, and the research agenda re-evaluated. A new Secretariat has been established.

Some Examples of Integrated Global Environmental Change Research

Without anticipating a revised IHDP research agenda, bridging the regional, the national and the global can especially be achieved at the international level of Global Change science through two complementary areas of activity: the operational and the methodological. The former should take into consideration the need to overcome geopolitical barriers, the latter should aim at developing better theories and at applying them.

The *operational level* should integrate national research needs and mechanisms (strongly represented through national committees) as well as regional (e.g., European, Inter-American, or Asia-Pacific such as ENRICH, IAI and APN, respectively) and international initiatives (e.g., START) into global environmental change research and capacity building, making sure that efforts are not duplicated and resources not wasted. Every opportunity must be grasped to improve the application of research results in the regional and international contexts. Goals, priorities, guidelines and quality control mechanisms should be suggested, and an overall coordination, archiving and synthesis of the results achieved. Key scientists and leading institutions should be identified, and partnerships or cosponsorships established with other relevant international research consortia.

The *methodological level* should be derived from existing theories and the nature of scientific problems. There is an absolute need to understand the interactions between regional and global processes as well as the links between the natural and social sciences. Some of these interactions are already addressed within specific themes concerning biogeochemical processes (International Geosphere-Biosphere Programme), physical processes (World Climate Research Programme) and socio-economic processes (International Human Dimensions Programme on Global Environmental Change). They embrace both topics and *methodologies*. For example:

LAND USE

A major project initiated by IGBP and IHDP deals with Land-Use and Land-Cover Change (LUCC). Human alterations of the landscape have had at least regional impacts over recorded history, but these are now global. They are the major causes of socio-economic reactions (e.g., soil degradation and groundwater depletion have consequences on population distribution, human health and commodity prices) and change on the biogeochemical and energy cycles, and thus climate (e.g., carbon dioxide releases from land clearing and biomass burning). Climate change in turn alters land use. LUCC will identify a range of land-use and land-cover dynamics, and better ways for land managers to deal with food security and sustainability issues. A Science Plan has been published [15] and an Open Science Meeting was held in Amsterdam (January 1996) to start the preparation of an implementation plan.

POPULATION AND FOOD PRODUCTION

Population growth drives socio-economic change and is intimately related to resource use, production and distribution patterns. Biological and cultural diversity as well as population dynamics are, therefore, intimately linked to sound policies addressing the need for sustainability aimed at approaching socio-environmental equilibria. A relevant example derived from research carried out by IGBP-IGAC (International Global Atmospheric Chemistry Project) has pointed out the fact that there are options to mitigate methane emissions from flooded rice fields which have a considerable impact on air chemistry and climate. A crucial reduction of global methane emissions will only require acceptance by the rice farmers of the world that it is in their interest to modify their current irrigation practices without loss of yields and gains in water and energy [17]. Another good example is the use of information on natural climate variability for the benefit of entire nations. The WCRP project caled TOGA (Tropical Oceans/Global Atmosphere) has, by giving skillful climate predictions on seasonal-to-interannual time scales for regions affected by ENSO (El Niño/Southern Oscillation), helped to turn a basic research result into operational use. Evidence suggests that Peru, for example, has adjusted its farming and fishing practices based on climate predictions and thus reduced economic losses during El Niño years. An IGBP project strongly illustrating social and economic impacts is LOICZ, the Land-Ocean Interactions in the Coastal Zone, which has a particular focus addressing socio-economic driving forces of and responses to global change for the 60 percent of the human population living along the coast.

ENERGY

Energy is related to population increase and the use and extraction of resources from ecosystems. People, resources and the environment have developed as a mutually interactive and co-evolving system. Increasing emissions of trace gases into the atmosphere have resulted from the industrial society's change of production processes and consumption patterns, based on the extensive use of fossil fuels, together with accelerated patterns of biomass burning from traditional agriculture and deforestation. To understand both the physical and the socio-economic systems, and the key interactions regulating the amount of energy used in relation to population density and consumption patterns, we need information from a broad spectrum of sources. Our sources include the past, the present, and prognostic biospheric and climatic models generated by researchers involved in the IGBP and WCRP, in order to synthesize the interactive natural and social mechanisms of change.

Global environmental research on energy requires transdisciplinary Earth system science of themes cutting across all fields of study from the IHDP [31], IGBP (mainly LUCC: Land-Use/Cover Change with the IHDP, IGAC: Atmospheric Chemistry and GAIM: Analyses and Modeling), and WCRP (GEWEX: Global Energy and Water Cycle and SPARC: Stratospheric Processes and their Role in Climate). Furthermore, the studies should provide decision makers with information to evaluate equity implications of abatement mechanisms in different socioeconomic contexts and scenarios, linking global energy and environmental policies to sustainable energy development.

FRESHWATER RESOURCES

This is probably the most pressing environmental research issue both on local and global scales. It is usually more obvious on regional or local levels where freshwater resource degradation is more conspicuous. However, there is a need to develop comparative research in terms of overall *vulnerability* to water shortage from communities to nations to continents, in order to find their differences between populations, socio-economic systems and within a variety of ecological settings. These studies can build on continental-scale hydrometeorological experiments of the Global Energy and Water Cycle Experiment (GEWEX) of the WCRP and on recent discussions to establish in the IGBP an inter-project initiative on "Hydrologic Transport and Water Resources" (associating BAHC, LOICZ and PAGES). The studies should suggest, among other potential results, ways in which early warning systems can be implemented in order to improve the responses of social systems to water security and drought. Water availability is, therefore, intimately linked with population, health, epidemic vulnerability, food production and environmental security. The recognition of all those topics are integrated in one of IHDP's working groups on "Global Environmental Change and Human Security" [20]. The studies will further need to link with ongoing initiatives dealing with global freshwater research and modeling within UN agencies (WMO, WHO, UNEP and FAO), and research institutes such as the Stockholm Environment Institute and the International Institute for Applied Systems Analysis.

Ultimately, the research agenda investigates the growing impact of anthropogenic environmental change on earth system processes supporting life. The enacted UN conventions on biodiversity and climate change have reacted to the disturbance of these processes and they need continuous scientific advice during their implementation. There is, therefore, a clear need to establish realistic priorities, to favor a problem-solving approach, and to extract policy-relevant results from which concerted strategies for mitigation and adaptation should emerge, and in which different partners are engaged, from the regional to the global scales. An example of such an effort is the recent launching of the International Research Institute for interannual climate prediction.

Understanding Global Environmental Change

In spite of the necessity for complementarity and pluralism in science, the search for better theories has to cultivate the right to be different, to disagree with established views, to produce new knowledge with originality. Therefore, on a methodological level, all major international research programs have to identify and to develop their own agendas. All scientific progress is built upon previous knowledge, which is best represented in each of the individual disciplines. In this sense there is never a new start, science and individuals who ultimately produce science are part of a continuum of social or disciplinary tradition, language and behavior. This should not be altered by forcing homogeneous attitudes, but existing environmental problems, and certainly a better understanding of their nature, are, in part, the result of a lack of discussion between the different disciplines. Complex social structures, including scientific research, must develop in face of the need to solve the fundamental problems of environmental change, and will ultimately be best addressed through an adequate degree of convergence and synergism among all three international Global Environmental Change programs.

Acknowledgments

I am grateful to Chris Rapley for his review of IGBP's research which I included with minor editing in the section on "Global Change and the IGBP." Some additional parts of the paper are also based in an article by myself, C. Rapley and H. Grassl published in the *Global Change Newsletter* **24**, December 1995, 23-25.

References

1. CIESIN (Consortium for International Earth Science Information Network), 1992: *Pathways of Understanding: The Interactions of Humanity and Global Environmental Change*. Ann Arbor, Michigan: CIESIN, 56 pp.

2. HDP (Human Dimensions Programme) Report No. 1, 1990: H.K. Jacobson and M.F. Price (eds.), *A Framework for Research on the Human Dimensions of Global Environmental Change*. Paris, France: HDP, 71 pp.

3. HDP Report No. 2, 1991: R. Worcester and S. Barnes (eds.), *Dynamics of Societal Learning about Global Environmental Change*. Paris, France: HDP, 158 pp.

4. HDP Report No. 3, 1992: J. Clarke and D.W. Rhind (eds.), *Population Data and Global Environmental Change*. Paris, France: HDP, 147 pp.

5. HDP Report No. 4, 1992: G. Yohe and K. Segerson (eds). *Economic Data and the Human Dimensions of Global Environmental Change: Creating a Data Support Proces for an Evolving Long-Term Research Program*. Paris, France: HDP, 70 pp.

6. HDP, 1994: *Work Plan 1994-1995*. Occasional Paper No. 6. Geneva, Switzerland: HDP, 45 pp.

7. IGBP (International Geosphere-Bisophere Programme), 1991: *Global Change System for Analysis, Research and Training (START)*. Report No. 5. Report of a Meeting at Bellagio, 3-7 December 1990, J.A. Eddy, T.F. Malone, J.J. McCarthy and T. Rosswall (eds). Stockholm, Sweden: IGBP, 40 pp.

8. IGBP Report No. 19, 1992: *The PAGES Project: Proposed Implementation Plans for Research Activities*, J.A. Eddy (ed.). Stockholm, Sweden: IGBP, 112 pp.

9. IGBP Report No. 21, 1992: *Global Change and Terrestrial Ecosystems: The Operational Plan*, W.L. Steffen, B.H. Walker, J.I. Ingram and G.W. Koch (eds.). Stockholm, Sweden: IGBP, 97 pp.

10. IGBP Report No. 27, 1993: *Biospheric Aspects of the Hydrological Cycle: The Operational Plan*, BAHC Core Project Office (ed.). Stockholm, Sweden: IGBP, 103 pp.

11. IGBP Report No. 28, 1994: *The IGBP in Action: The Work Plan 1994-1998*. Stockholm, Sweden: IGBP, 151 pp.

12. IGBP Report No. 30, 1994: *IGBP Global Modelling and Data Activities, 1994-1988*. Strategy and Implementation Plans for Global Analysis, Interpretation and Modelling (GAIM) and the IGBP Data and Information System (IGBP-DIS). Stockholm, Sweden: IGBP, 86 pp.

13. IGBP Report No. 32, 1994: *International Global Atmospheric Chemistry (IGAC) Project: The Operational Plan*. Stockholm, Sweden: IGBP, 134 pp.

14. IGBP Report No. 33, 1995: *Land-Ocean Interactions in the Coastal Zone: Implementation Plan*. J.C. Pernetta and J.D. Milliman (eds.). Stockholm, Sweden: IGBP, 215 pp.

15. IGBP Report No. 35, 1995: *Land-Use and Land-Cover Change: Science/Research Plan*. B.L. Turner II, D. Skole, S. Sanderson, G. Fischer, L. Fresco, and R. Leemans (eds.). Stockholm, Sweden/Geneva, Switzerland: IGBP/IHDP, 132 pp.

16. IGBP, 1995: *IGBP Directory 1995*. Stockholm, Sweden: IGBP, 41-52.

17. IGBP, 1995: *IGBP Newsletter*, **22**, 4-5.

18. Kasperson, J., Kasperson R., and Turner II, B.L. (eds.), 1995: *Regions at Risk: Comparisons of Threatened Environment*. Tokyo, Japan: United Nations University Press.

19. Kosinski, L. (ed.), 1996: *Issues in Global Change Research: Problems, Data and Programmes*. HDP Report No. 6. Geneva, Switzerland: HDP, 113 pp.

20. Lonergan, S., 1996: Environmental security. *Scoping Report*. Geneva, Switzerland: IHDP, 38 pp.

21. Luterbacher, U., 1996: Demographic and social dimensions of resource use. *Progress Report*. Geneva, Switzerland: IHDP, 24 pp.

22. McMichael, A.J., 1996: Human health and environment. *Scoping Report*. Geneva, Switzerland: IHDP, 12 pp.

23. Pawlik, K., 1996: *PAGES Work Plan*. Geneva, Switzerland: IHDP, n.p.

24. Price, M., 1990: *Report of the International Social Science Council Scientific Symposium on the Human Dimensions of Global Environmental Change*. Occasional Paper No. 1. Geneva, Switzerland: HDP.

25. Price, M., 1994: *Options for EC-Level Research Activities on the Human Dimensions of Global Change*. DG/XII/D. Brussels, Belgium: European Commission, 69 pp.

26. Redclift, M., and T. Benton (eds.), 1994: *Social Theory and the Global Environment*. London: Routledge, 271 pp.

27. Stycos, J.M. and S. Simões, 1996: *Global Environmental Survey: Work Plan*. Geneva, Switzerland, IHDP.

28. Underdal, A., and O. Young, 1996: *Institutions: Scoping Report*. Geneva, Switzerland: IHDP, 29 pp.

29. Williamson, P., 1992: *Global Change: Reducing Uncertainties*. Booklet. Stockholm, Sweden: IGBP, 40 pp.

30. Vellinga, P., 1996: *Industrial Transformations: Progress Report*. Geneva, Switzerland: IHDP, 29 pp.

31. Yohe, G., 1996: *Energy Production and Consumption: Progress Report*. Geneva, Switzerland, IHDP, 11 pp.

32. Zang, L., 1996: *Trade and Environment: Scoping Report*. Geneva, Switzerland, IHDP, 5 pp.

DEVELOPMENT OF THE CASPIAN-ARAL SEAS PROGRAM: AN ICWC PERSPECTIVE

VICTOR A. DUKHOVNY and VADIM I. SOKOLOV
International Coordination Water Commission (ICWC) of the Aral Sea Basin
Tashkent, Uzbekistan

The main purpose of this paper is to generate a preliminary discussion of the possibilities of supplying the Aral Sea basin with water from outside regions via long-range water transfers. One such alternative may be a water transfer from the Caspian Sea basin. A few different variants of this transfer have been proposed, but we are not going to discuss them in detail. Instead, we discuss the possibility of such project in principle.

First of all, it is necessary to present the present situation in the Aral Sea basin, and from this it will be clear that supplying the basin with water from outside in the nearest future is the objective of this proposal.

The Aral Sea and its basin have gained notoriety, as a result of interest of the global community in the degradation of bodies of water; the Aral Sea is one of the worst examples. Until 1960, the Sea covered an area of about 66,000 km^2 and had a volume of more than 1,000 km^3. It has been fed by inflows from the Amudarya and Syrdarya which used to total approximately 47 km^3 per year. With these inflows, the level of the Sea was maintained at between 51.0 and 53.0 m; there was a stable water balance in the Aral Sea.

Between 1960 and 1990, an increase in water use in the Basin from 63 km^3/year altered the balance of the Aral Sea (see Table 1). by 1990, the total annual inflows from the Amudarya and Syrdarya to the Sea had declined to 9.0 km^3 and the level of the Sea had dropped 15 m to the current level of about 38 m. The Sea had lost about 50% of its surface area since 1960.

M.H. Glantz and I.S. Zonn (eds.),
Scientific, Environmental, and Political Issues in the Circum-Caspian Region, 185-190.
© 1997 *Kluwer Academic Publishers. Printed in the Netherlands.*

TABLE 1. The dynamics of water resources distribution in the Aral Sea basin

Years	GNP	Including		Population	Water Consumption		Water Resources	Intake Abroad	Runoff Losses	Into Aral Sea
	Bill. Rubl	Industry	Agriculture	Million	Industry	Agriculture	km^3	km^3	km^3	km^3
		Bill. Rubl	Bill. Rubl		km^3	km^3				
1960	16.1	6.6	9.5/4.5*)	14.2	2	61	120	1	9	47
1975	40.6	23.8	16.8/6.8*)	24.8	5.2	80	120	1.5	9.3	24
1990	74	47	27/7.2*)	36.4	7	91	120	2.2	10.8	9

*) irrigated area, million ha

The total renewable water resources (surface and underground) in the Aral Sea Basin are about 120 km^3/year. In 1960 there was a population of 14.2 million and about 4.5 million hectares of irrigated land in the basin. The total GNP was about 16.1 billion Rubles (in 1983 prices) (see Table 1). By 1990, the population had increased to 36.4 million, the irrigated area to 7.2 million ha, and the GNP to 74 billion Rubles.

Taking into account the new political, economic and social realities, and recognizing the severity of environmental concerns, the Heads of the five Central Asian States — Kazakstan, Kyrgyzstan, Tajikistan, Turkmenistan and Uzbekistan — met in January 1994 in Nukus, Karakalpakstan, and approved an Action Plan for the improvement of the environmental situation in the basin. The plan also addressed social and economic development for the following three to five years. A regional water management strategy is a centerpiece of the Action Plan. One of the principal objectives of a regional water management strategy is the conservation and a more efficient use of water, especially in relation to irrigated farming. A gradual transition from a system of "norms and quotas" to "demand management" using economic and financial incentives is unavoidable, especially in view of land-tenure changes. It is estimated that about 20% of the water currently used for irrigation can be saved for other purposes. In principal the water conservation program would include:

- Development of basin-wide norms of water use in all branches of the economy, especially in irrigated agriculture.
- Assessment of the real potential productivity of water and land.
- Development of criteria for the evaluation of water-use efficiency.
- Introduction of payment for water.
- Control of water losses from river channels.
- An agreement among the states about national water limits.
- Introduction of advanced water-use technologies.
- Introduction of a water conservation advisory (extension) service.

As a result, all five states agreed to the future normative demands for water at the ICWC meeting, that was held in Charjou (Turkmenistan) in January 1996 (see Table 2). For irrigation water, consumption will decrease in the Aral Sea basin from the current level of 139.5 km^3/year to 103.7 km^3/year in 2010, and the total demands for water by all economic sectors will decrease from 152.0 km^3 to 125.2 km^3 in the same period.

As shown in Table 3, a comparison of the actual and calculated balances for 1994 indicates that to supply all of the existing irrigation system at a level compatible with the present "norms" of specific water-use per hectare a total volume of 151.8 km^3 instead of 111.7 km^3 would have been required. Assuming that the water inflow to the Aral Sea was at the same level as was actually recorded in 1994, the calculated balance for 1994 shows a water deficit in the basin of 36.1 km^3.

TABLE 2. Comparison of current and future normative demands for water by economic sectors of Aral Sea basin countries (millions m^3)

Country	Calculated level	Economic sectors						Total
		Dom. + Drink	Rural water supply	Industry	Fish	Irrigation	Other	
1	2	3	4	5	6	7	8	9
Kazakstan	1994	215.7	137.5	516.7	234.6	12,089.1	491.2	13,684.8
	2010	383.9	229	474.1	341	10,935.3	600	12,963.3
Kyrgyzstan	1995	28.9	70	74.8	—	9790	—	9963.7
	2010	69.9	80	137.1	—	9170	—	9457
Tajikistan	1990	487	696	594	—	11,764	374	13,915
	2010	770	550	1500	500	10,380	600	14,300
Turkmeni-stan	1995	620	210	730	40	31,330	—	32,930
	2010	1100	270	2250	400	25,225	—	29,245
Uzbekistan	1995	3200	1000	1450	1260	74,600	—	81,510
	2010	5850	1635	1460	2255	48,000	—	59,200
Total for Aral Sea Basin	Current	4551.6	2113.5	3365.5	1534.6	139,573.1	865.2	152,003.5
	2010	8173.8	2764	5821.2	3496	103,710.3	1200	125,165.3

TABLE 3. Water balance in the Aral Sea Basin (km^3)

	1994 Actual	1994 Calculated	2010 Calculated
	Total Aral Sea Basin	Total Aral Sea Basin	Total Aral Sea Basin
RESOURCES			
Surface water resources	121.5	121.5	115.6
Groundwater recharge	14.0	14.0	18.0
Return flow	45.9	45.9	31.4
Total Resources	**181.4**	**181.4**	**165.0**
USES			
Unaccounted losses from river channels	10.2	10.2	10.0
Surface flow losses due to groundwater withdrawal	7.5	7.5	10.6
Unused discharge of vertical drainage water	2.6	2.6	4.0
Summary withdrawal by all economic sectors	111.7	151.8	125.2
Diversion of return flow to Aral Sea and natural depressions	16.4	16.4	9.5
Water withdrawal by Afghanistan	2.0	2.0	5.0
Discharge into the Aral Sea by the main streams	26.8	26.8	19.0
Total uses	**177.2**	**217.5**	**183.3**
BALANCE	**+4.2**	**-36.1**	**-18.3**

Table 3 also presents the calculated prospective balance for the year 2010. Still, even under the lower-demand scenario, the shortage of water in the Basin is on the order of 18.3 km^3 (assuming Aral Sea inflow at the level of 19 km^3 per year, a level which cannot sustain the Sea even at its present level).

Thus, it is clear that the Aral Sea basin will require about 18.0 km^3 per year of water from outside sources at the level of 2010 (see Table 3). As mentioned above, one of the alternative sources for water is the transfer of water from the Caspian Sea basin. The "last word" on this subject, however, belongs to Russia and other Caspian countries. If Caspian countries and states of Central Asia can find mutual understanding on this problem, it could be a good chance to apply to the World Bank and other international donors to support research on these issues.

We can, for example, revise the old idea of taking water from the Volga River through the Ural River and then to transfer about 10 to 15 km^3/year to the Aral Sea region.

IRANIAN PERSPECTIVE ON ENVIRONMENTAL PROBLEMS IN THE CASPIAN SEA REGION

AMIR BADAKHSHAN
The University of Calgary
Calgary, Alberta, Canada

JALAL SHAYEGAN
Sharif University of Technology
Tehran, Iran

Introduction

The Caspian Sea, the largest lake in the world, is located north of Iran and to the south of Russia. The lake is divided into three recognized regions:

- **North region**, with an average depth of about 62 meters, covers about 28% of the lake's total area but less than 1% of total volume.

- **Central region**, with an average depth of about 176 meters, covers about 36% of the area and 33% of the total volume.

- **Southern region**, with an average depth of about 330 meters, covers 36% of the area but about 66% of the volume. In some locations of this region, the water depth is nearly 1,000 meters.

Iran's share of the lake is: area 10%, catchment basin area 7%, sea coastline 14%, annual settling solids 3%, annual fishery (1000 metric tons) 2%, annual water input 5%.

The Caspian Sea receives waters from the Volga River, which drains approximately 20% of the European land area. In addition, it also receives Ural River, Kura River and numerous other fresh water river inputs, including those of Iranian rivers.

Four hundred fish species are endemic to Caspian waters and some of these, notably the sturgeon, are of major economic importance for all five countries surrounding the sea. Bird life is prolific throughout the year with large population expansion during the migration seasons, when many birds return to the extensive

M.H. Glantz and I.S. Zonn (eds.),
Scientific, Environmental, and Political Issues in the Circum-Caspian Region, 191-209.
© 1997 *Kluwer Academic Publishers. Printed in the Netherlands.*

deltas, shallows and other wetlands. Many individual species are threatened by impacts ranging from over-exploitation, to habitat destruction, to polluted environments.

The Caspian basin is rich in hydrocarbon deposits (see Figs. 1-12). It has an estimated proven and probable reserve of about 50 billion barrels (20 x 10^9 metric tons as suggested by one of the Russian representatives in the workshop). The present production rate is likely to increase rapidly as a result of exploration activities which are currently under way. The importance of Caspian Sea livelihood to the well-being of its neighboring countries and its delicate environment calls for a strong commitment for its protection and full cooperation among the riparian governments.

Pollution of the Caspian Sea (Manmade)

The sources of pollution of the Caspian Sea by the neighboring countries can be divided into two main categories:

1. Activities on shore that create wastes which are then disposed of in the sea.
2. Activities off shore.

Activities on shore: Pollution created by activities on shore is widespread and important, and each country should be responsible for its reduction.

Activities off shore: Sea transportation of commercial and oil tanker fleets, and petroleum exploration and production are the primary contributors to off-shore pollutants.

A third very important category, namely radioactive pollution, can also be added, which could belong to either category (on-shore, atomic power station wastes; and off-shore, nuclear power submarines).

It is expected that there will be numerous major activities involving the exploration, production and transportation of petroleum in this sea in the near future. Therefore, this chapter describes the potential for polluting the sea at different stages of these activities and tries to introduce ways either to prevent or to reduce their damaging impacts.

Environmental Protection in Offshore Exploration and Production Operations

INTRODUCTION

The international petroleum industry, which began in Pennsylvania in 1889, quickly swept across the continents of the world. In the 1890s, production was established under the Pacific Ocean through directional wells, drilled on shore or from piers extending out from the beach.

In the 1890s wells were drilled in the shallow water of Lake Maracaibo, Venezuela, from wooden platforms. Similarly, at an early date, offshore production was established from platforms and piers in the Caspian Sea at Baku, Azerbaijan. The first drilling platform beyond the sight of the land (off the coast of Louisiana) was constructed in about 1948. From those early beginnings, the offshore industry as we know it today has developed.

Here we discuss the various steps that petroleum activities should undertake in order to avoid, or at least minimize, unnecessary adverse environmental impacts on the Caspian Sea.

ENVIRONMENTAL EFFECTS

Studies have shown that the quantity of hydrocarbons entering the marine environment from the natural seepage of oil in the sea ranges from 200,000 metric tons to perhaps six million metric tons/year, with the best estimate being 600,000 tons/year.

Information concerning spills is not readily available, and assessment of the level of present oil spills in the Caspian Sea must be included in any regional cooperation. The environmental impact of marine exploration and production operations may be classified as follows:

Geophysical Operations

The beginning of oil operations in an offshore area is marked by the arrival of a geophysical survey crew. This crew procures data necessary to prepare an interpretation of the sub-sea strata.

To create the sound waves necessary for the operation, heavily charged explosives are used. The explosives are a source of pollution in the water, which results in changes to marine life. The operations may also affect fishing activities. Air guns, which provide shear impulses from compressed air, should replace explosives whenever possible.

Drilling Operations

Several aspects of drilling operations have potential adverse impacts on the environment and, hence, are of concern.

1. The effect of the oil rig upon the immediate environment, merely by its location upon the well as well as drilling operations themselves.
2. The possibility of blowout occurring during the drilling operation which results in pollution by fluids from within the earth.

To minimize the impact of pollution on the environment, the following steps should be taken:

1. Strict implementation of safety procedures.
2. Proper location of the drilling rig with regard to the sea lanes.
3. Proper survey of the sea bottom for important information relative to the anchoring of a semi-submersible or in locating jack-up rigs.
4. Collection of seismic information necessary for the proper selection of well location.
5. Proper selection of rig materials, and the self propelling of the rigs. Provision for the dynamic positioning of the drilling vessel without resorting to the use of anchors.

6. Minimizing blowout by:

    a) Proper design of a casing scheme.
    b) Proper selection, installation, testing and maintenance of blowout prevention equipment.
    c) Proper selection and maintenance of the mud program.
    d) Proper training of rig personnel.
    e) Establishment and adherence to appropriate emergency procedures.
    f) Addition of motion compensators to floating rigs to increase drilling efficiency by eliminating the relevant movements between drill string and the bottom of the well.
    g) Installation of various sensor — and mud — volume and pressure monitoring devices, which would recognize parameters for an early warning of a possible blowout.

Drilling Fluid

Contaminated cutting and drilling fluids (water- and oil-based) must not be released overboard, unless the cuttings are properly cleaned.

Flaring of hydrocarbons during well-testing

Efficient high-volume burners must be used for the proper burning of the temporary production of the wells during testing and to prevent liquid hydrocarbons from entering the sea. Directional drilling from a well-equipped platform, as opposed to single-well drilling, reduces additional pollution.

Offshore Loading

The transportation of crude oil from an offshore oil field, by way of a conventional pipeline carrying partially degassed crude, has several advantages over offshore loading of degassed crude into tankers (e.g., pipeline operations are not affected by weather). In this case, the production rate is continuous and operational. Furthermore, offshore loading also has the following adverse environmental impacts:

a) Chance of spillage;
b) Hazards associated with tanker movement in the proximity of the platform; and
c) Potential contamination due to the discharge of ballast water.

Pipelines

Although offshore pipelines have positive value, they also have the potential for substantial impact on the environment if proper care is not taken for their design and operation. Therefore, the following should be considered for offshore pipelines:

a) Pipes should meet stringent metallurgical and dimensional quality control;
b) Each length of pipe should have an anti-corrosion coating (prevention of mechanical damage);
c) When necessary, the pipeline should be buried in the seabed for extra protection; and
d) Sensors should be installed on pipelines to monitor pressure drops due to breakage. These sensors should be able to shut off the wells, when necessary.

Offshore Production Facilities

Proper construction of steel structures for production facilities needs sound design parameters. Holding the structures in place could be accomplished by driving pilings through through hollow tubular legs which penetrate deeply into the sea floor or by the use of gravity structures which are maintained by their own weight. New technologies for sub-sea construction are continually being developed. Besides specific information for production structures needed for the design of installations, information on long-term weather patterns should be available prior to the design of offshore production facilities.

- Various sensors and controls should be installed on the platform, on pressurized vessels, and water-separator facilities.
- Offshore platforms should also be equipped with flame, smoke and natural gas sensors which can automatically shutdown the operation.
- Proper fire-fighting equipments should be installed.

Combatting offshore blowouts

To combat a blowout at sea, a regional oil-spill contingency plan should be designed to be implemented immediately upon the observance of an oil spill and the following steps should be taken:

1. List jobs that must be done, indicating priorities.
2. Prior to an oil spill, assign such jobs to appropriate personnel giving them authority.
3. Establish communication patterns to assure coordination.
4. Assure the availability of materials that might be needed.

Various government agencies should be promptly notified and a proper plan of attack should be developed, as shown below:

1. Obtain spray equipment to cool the vital jacket section of the platform to preserve its integrity (blowouts at sea must be put on fire).
2. Allow as much as possible of the well effluent to burn in order to protect the water environment from pollution.
3. Secure all necessary equipment to confine and recover any oil slick that might develop.
4. Secure drilling rig(s) and commence drilling a relief hole for a “bottom kill” on all wells.
5. Arrange a work platform adjacent to the burning structure to provide a stable base for effecting a “top kill” or, if necessary, from which to effect repairs after shut-off.
6. Consult with experienced fire-fighting authorities.
7. Call the company’s (or region’s) booms and skimmer facilities along with a tanker barge to recover and transport oil from the surface of the water.

OIL SPILL CLEAN UP AND CONTROL

For any oil spill, four principles must be applied:

1. Limit the size of the spill;

2. Contain the spill;
3. Recover the spilled oil, whenever possible; and
4. Minimize environmental damage.

Two methods are available for containment: The use of booms and the use of surface tension modifiers. Both have limitations, especially in bad weather. Once a spill is contained, the oil can be recovered by several types of skimmers. For the open sea, the application of dispersants, which accelerate biodegradation of the residual oil, is the best approach.

INTRA-INDUSTRY COOPERATION

An organization such as the Clean Gulf Associate (CGA), formed in the Gulf of Mexico, should be created for the Caspian Sea with all of the oil companies operating in the sea as members.

The goal of the organization would be to:

1. Provide oil spill containment and clean-up capability;
2. Maintain equipment at strategic locations in a state of 24-hour readiness; and
3. Upgrade their equipment with technical advances.

Other similar organizations:

- Gulf Area Companies Mutual Aid Organization (Middle East);
- Others exist in Venezuela, Alaska and Japan.

In the North Sea area each company has emergency action groups, e.g., United Kingdom Offshore Operations Association (UKOOA) and the Netherlands Oil and Gas Exploration and Production Association (NOSEPA). Organizations exist worldwide to handle the compensation of affected parties on behalf of their member parties. These organizations may be good examples on which to base a Caspian Sea organization for the same purpose.

IPIECA (International Petroleum Industry Environmental Conservation Association) membership may also be of help to local companies. Through these organizations the exploration and production segment of the industry seeks to consult and cooperate with relevant authorities in framing clearly defined, practical and non-discriminatory environmental regulations.

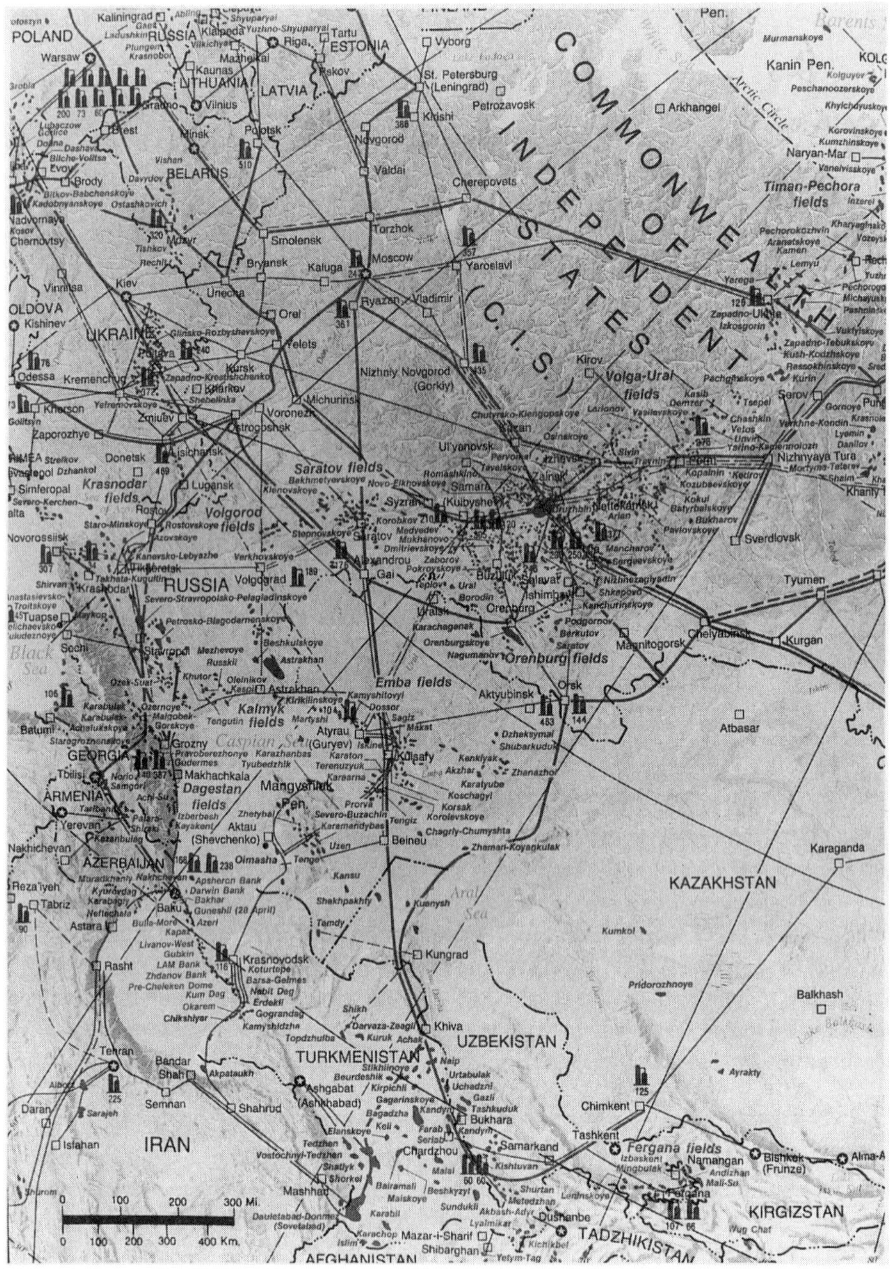

Fig. 1. Caspian Sea region oil map

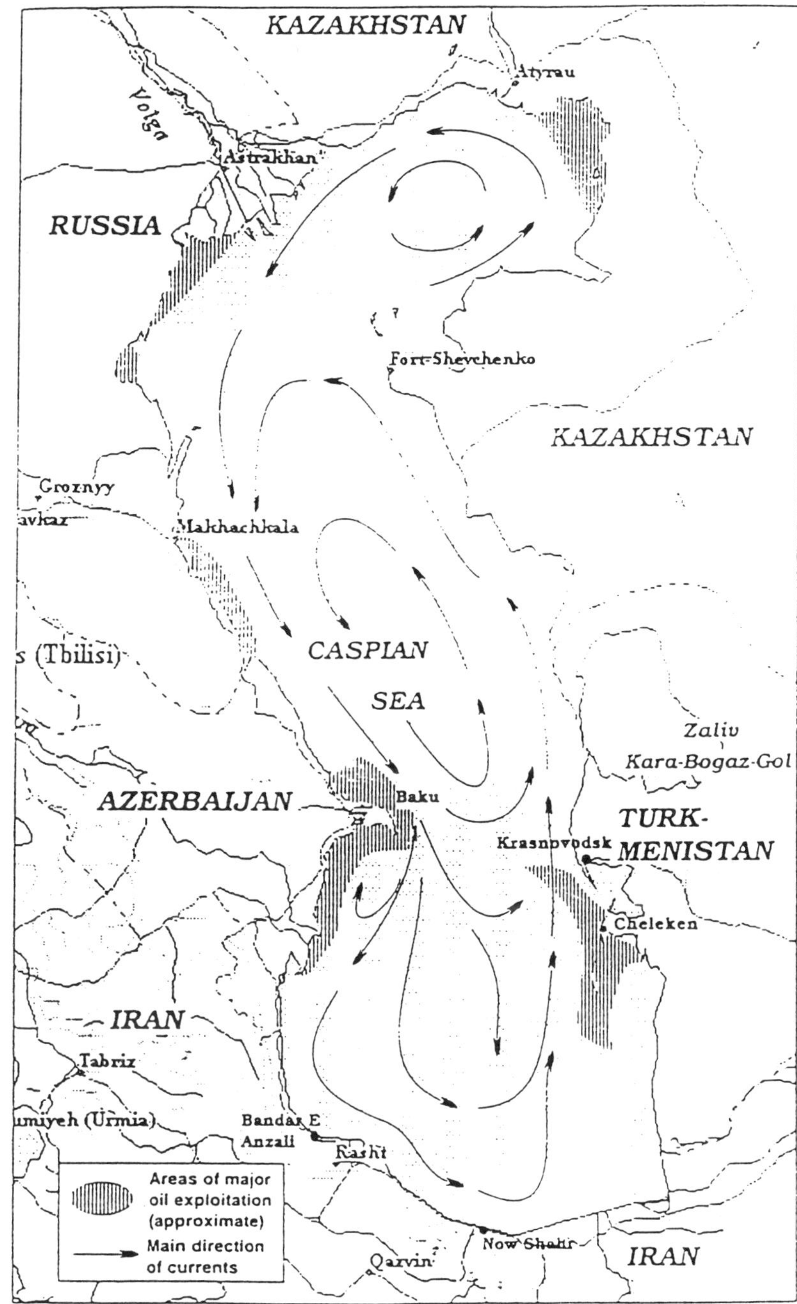

Fig. 2. Areas of major oil exploitation

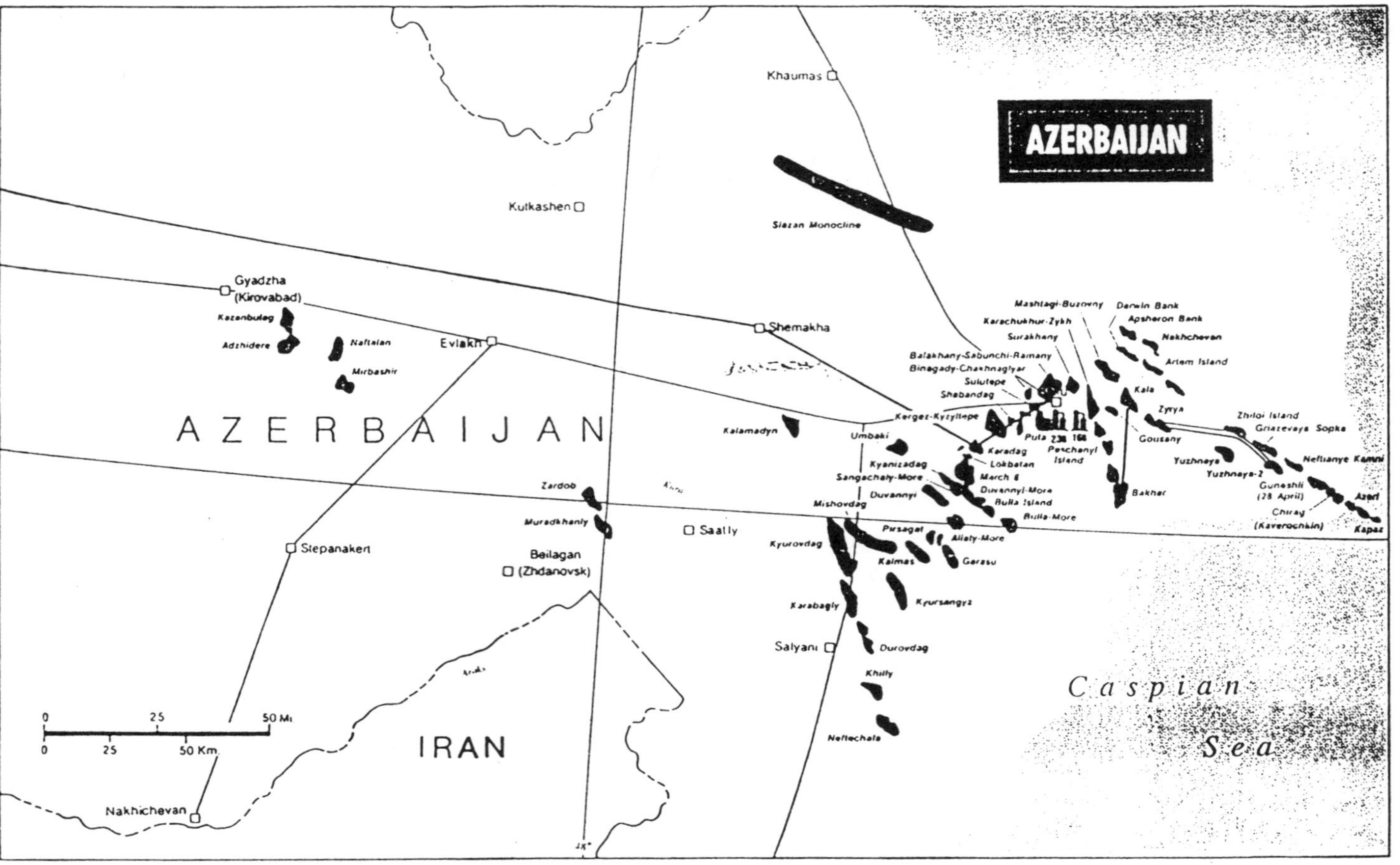

Fig. 3. Azerbaijan oil and gas activities

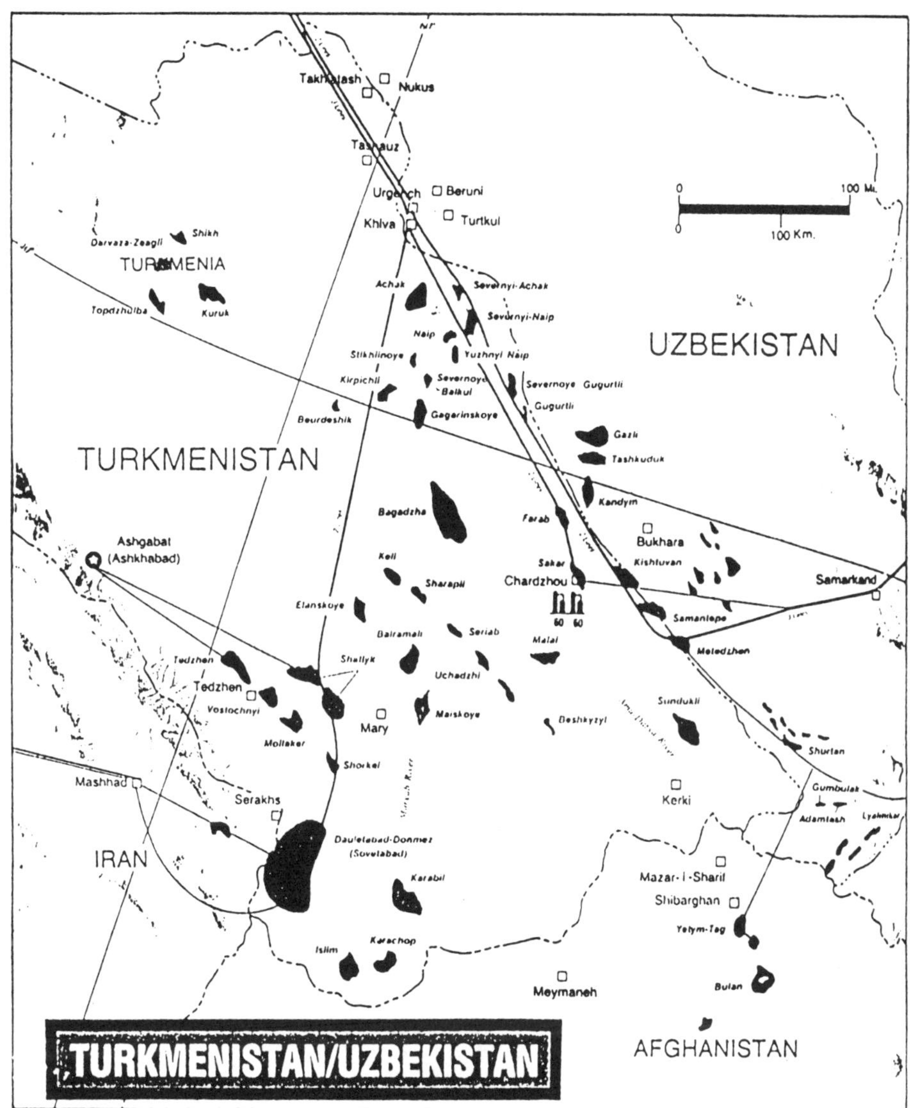

Fig. 4. Turkmenistan/Uzbekistan oil and gas activities

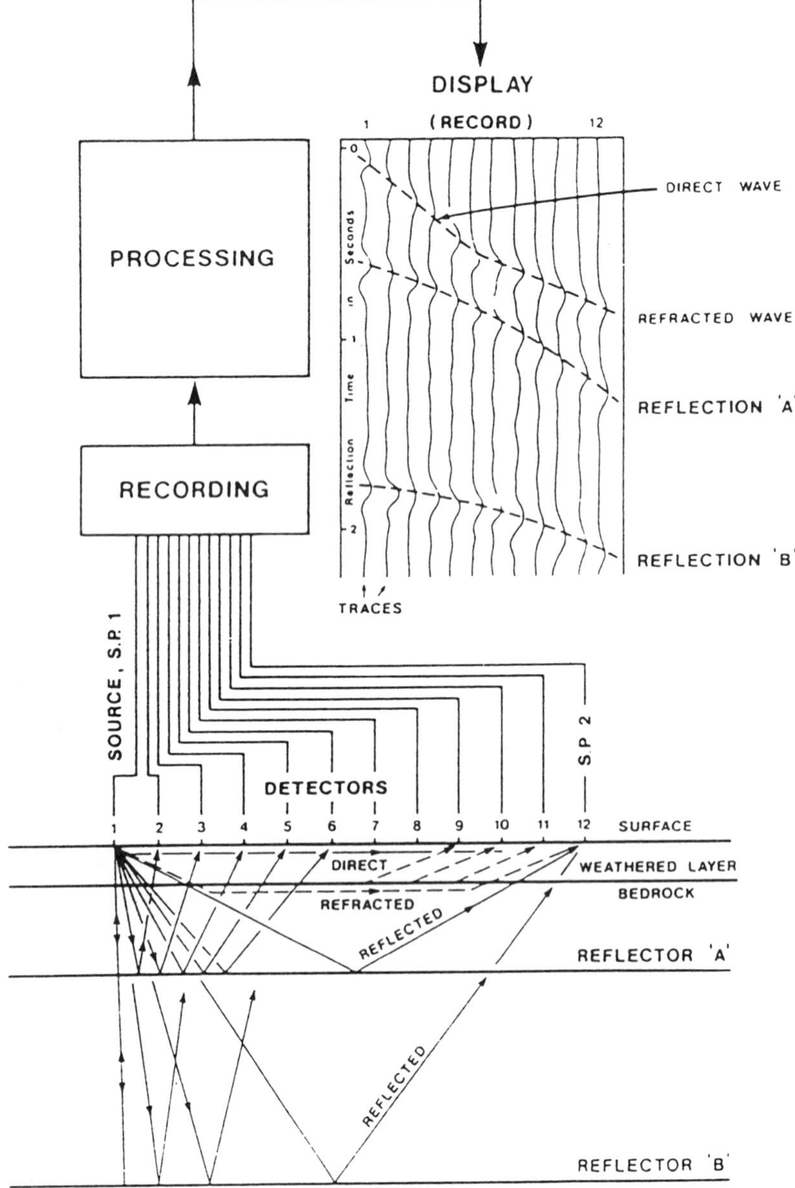

Fig. 5. Schematic diagram of traditional seismic reflection survey

Fig. 6. Fixed multi-well platform

Fig. 7. Jack-up platform

Fig. 8. Semi-submersible platform

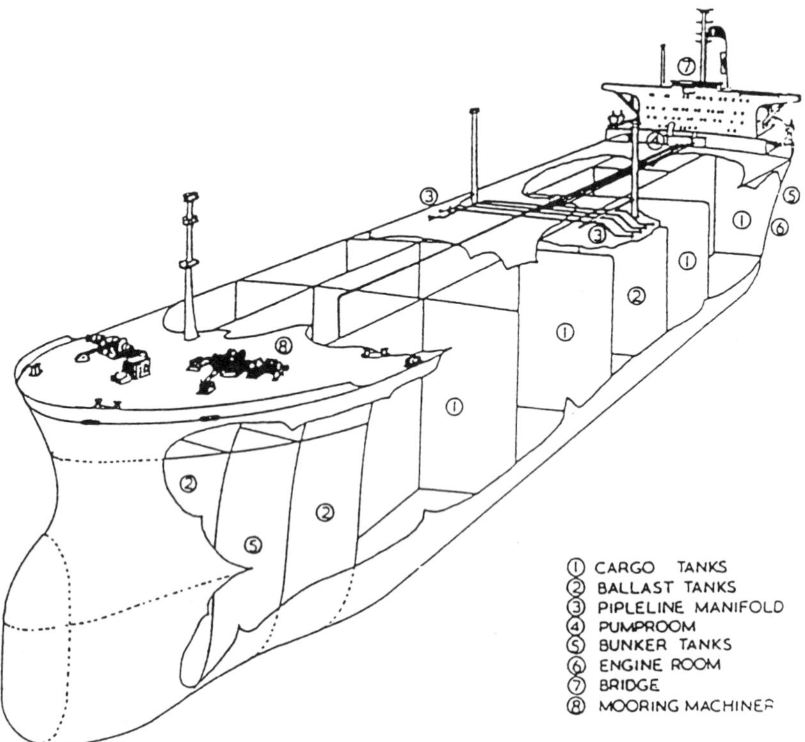

Fig. 9. Typical layout of a 220, 000 d wt tanker

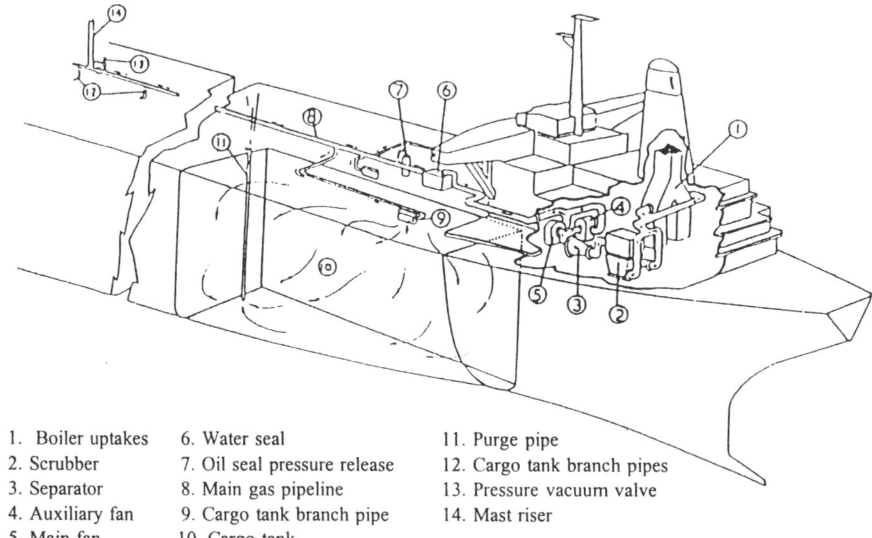

Fig. 10. Inert gas system in an oil tanker

Fig. 11. Eleo Maersk, very large crude-oil carrier in a double hall tanker

Fig. 12. Oil spill handling

Further Reading

1. Hobson, G.D. (Ed.), 1984: *Modern Petroleum Technology*. Fifth Edition, Volumes 1 and 2, Institute of Petroleum. New York: John Wiley and Sons.

2. Anon., 1994: *International Petroleum Encyclopedia*. Tulsa, Oklahoma: Pen Well Publishing Company.

3. Petroleum and Environmental Conservation, 1975: *Symposium Proceedings*. Report of the proceedings at the IPIECA Symposium, "Petroleum and Environmental Conservation," held in Tehran, Iran, 7-9 April 1975. London, England: International Petroleum Industry Environmental Conservation Association (IPIECA).

4. Badakhshan, A., 1996: Lecture Notes, ENCH 645, Environmental Engineering. Unpublished. Available from Department of Chemical and Petroleum Engineering, The University of Calgary, Calgary, Alberta, Canada T2N 1N4.

V. Examples and Lessons from Other Regions

AN OVERVIEW OF THE ENVIRONMENTAL ISSUES IN THE BLACK SEA REGION

M. ÖZTÜRK and F. ÖZDEMİR
Ege University
Science Faculty
Biology Dept.
Izmir, Turkey

E. YÜCEL
Anadolu University
Science Faculty
Biology Dept.
Eskişehir, Turkey

Introduction

Bulgaria, Georgia, Rumania, the Russian Federation, Turkey and Ukraine share a common heritage, namely the Black Sea, which has been a centerpoint for many civilizations. This sea has served each of these countries as a backbone in their economies, but has at the same time been used as a waste receptacle by them. Both industrial and domestic effluent, as well as heavy tanker traffic, have severely polluted the Black Sea. The load of pollutants has increased tremendously by way of 34 large rivers that enter the sea from Asia and Europe.

Eutrophication is a common sight on its coasts. Fish stocks on the Turkish side have declined, as has happened on other sides as well. Most of the beaches are not usable. The Sea of Azov is dying and the sea resort of Sochi has no longer been a famous health resort after the Chernobyl accident. The landlocked Black Sea is facing a catastrophic situation and should be considered as one of the region's major environmental problems. All interested parties need to join hands to extend their efforts to save this water body.

This chapter evaluates the Black Sea situation in general, with suggestions for a sustainable use of ecological and biological processes of this unique marine ecosystem.

M.H. Glantz and I.S. Zonn (eds.),
Scientific, Environmental, and Political Issues in the Circum-Caspian Region, 213-226.
© 1997 *Kluwer Academic Publishers. Printed in the Netherlands.*

The Black Sea is a part of the chain of marine ecosystems which constitute 3/4 of our planet. This semi-enclosed sea [41], is connected weakly with the Mediterranean Sea through the nearly 50m-deep Straits of Bosphorus and Dardanelles, but is extremely poor in islands as compared to the latter and possesses a 50 cm higher water level. Some investigators accept it as relic of an old ocean; others see it as a depression area formed by tectonic plate activity.

According to the information put forth by Demirsoy [9], it was a big shallow lake like the Sarmatic inland sea with brackish water characteristics from Hungary to Turkmenistan until the end of the tertiary period and, together with the Aral and Caspian Seas, is a remnant of Tethys. During the Miocene, Tethys separated into northern and southern parts, forming the Paratethys including the Aral, Caspian, and Black Sea areas which were eastern extensions of it. During the mid-Pliocene, the Black Sea separated from the Aral-Caspian. Its present shape is a result of tectonic activities which occurred during the post-tertiary period. Its present depth conditions evolved some 2,500 years ago, varying in depth between 15 m as in the Sea of Azov up to 2,245 m in the north of Ayancik, Turkey. The waters of the Mediterranean have entered this sea three times, affecting its hydrology and biological diversity. The Black Sea exhibits miniature oceanic characteristics. Its maximum length is 1,200 km from east to west from Burgas, Bulgaria to Ponti, Georgia and its width is 600 km from Ereğli, Turkey to Odessa, Ukraine. The total surface area is 459,064 km^2. A large number of water inlets enter the sea from the bordering countries, of which 34 are large ones which transport a heavy load of pollutants to the sea.

A beautiful landscape around the sea, good recreational areas and heavy industrialization have been major forces for the demographic "explosion" along the coast. As a result of these, the input of domestic as well as industrial effluent into the Black Sea increased considerably. The sea is facing a disastrous future because of these effluent: large amounts of solid wastes, pesticides, fertilizers (e.g., nitrogen, phosphorus and synthetic fertilizers), toxic industrial chemical wastes, nuclear residues, petroleum and bilge water. The environmental problems created by the pollutants from these sources and from the Danube and the Rhine are at present severe. The circulation of currents in the Black Sea brings this pollution load toward the Turkish coast and then passes it on to the Mediterranean Sea through the straits [2]. The aim of this review is to evaluate this situation and compare it with the Caspian Sea.

Demographic Developments

The coastline of the Black Sea encompasses nearly 30 major cities in six riparian states with densely populated settlements. In the Russian Federation alone 6-700,000 migrants have come to the coast from different areas in order to work [45]. Irrigated areas around the Sea of Azov are over 500,000 km^2 and the one around the

Danube, 417,000 km^2; this led to population growth in these areas. A perusal of Turkish census reports [12] reveals that, the population in 1945 was 4.5 million, increased to 6 million in 1960, and to 8 million in 1990 (Fig. 1). The demographic growth rate was highest in Samsun followed by Zonguldak, Trabzon and Giresun. The reasons for this were as follows: the availability of agricultural land for rice and tobacco cultivation in Samsun, coal excavation in Zonguldak; copper mining, and hazelnut, tea and chestnut production in the eastern part. Turkey dominates the hazelnut market in the world, producing 305,000 metric tons a year [12]. The annual tea and chestnut production is around 120,000 and 80,000 metric tons, respectively.

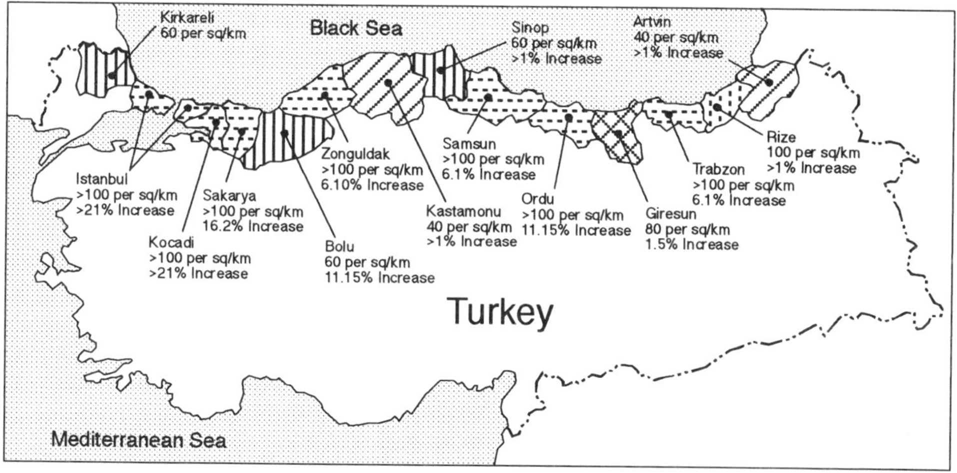

Fig. 1. Change of population during 1985-90 along the Turkish Black Sea Coast

In a walk at random along a 100-meter strip of the Turkish coast, one can collect over 100 plastic bags or bottles, nearly 250 lids or wrappers and over 50 empty metal cans for beer and soft drinks. A 100% increase in the population in the coastal parts of Turkey during a period of five decades has doubled domestic-waste production as well. There are very few efficient sewage treatment plants. The situation has been made worse by the apparent condoning of the belief accepted by bordering countries that the Black Sea is a receptacle for waste.

Oil Spills

World marine trade has increased by 2.4% and reached a level of 4.3 billion metric tons in early 1994 in spite of the fact that overall economic development did not follow a similar favorable trend. The number of ships above 300 Grt (gross tonnes)

has increased from 33,500 to 35,158 [10] of which 1,776 belonged to the Russian Federation, 739 to Turkey, 508 to Ukraine, 248 to Rumania and 126 to Bulgaria; including respectively, 275, 87, 33, 19 and 16 tankers. According to 1994 statistics, the total number of ships in Turkey increased from 3,574 (1986) to 4,787 (1993); these include 1500 ships of lower tonnage and 400 yachts. An additional annual traffic load exists of over 2,600 yachts from neighboring countries during summertime as well as oil tankers that cross the Black Sea and the Mediterranean Sea [27]. During transportation activities, 385,000 metric tons of waste are dumped into world marine ecosystems from the ships and this amount is expected to increase as traffic does [52].

Oil production in the world has crossed the level of 2.2 billion metric tons together with 0.44 billion metric tons coming from marine exploration. The increase in production during the last 4 decades has been 10 fold. Part of this is transported by sea by 4,700 tankers, of which 700 are supertankers; 75-85% of it is crude oil [52]. During this activity, 0.95-2.62 million metric tons of petroleum enter the marine environment annually [26]. Petroleum contributions to global environment pollution suggest that transportation-related activities and accidents account for at least one-third of the total oceanic oil pollution. During the last 2 decades more than 200 marine accidents took place in the Istanbul Straits alone, including the big Rumanian tanker accident in 1979, which spilled 64,000 tons of oil and caused the death of 43 persons. In 1991, due to a collision of Lebanese ship with a Philippine tanker, 22,000 sheep drowned; in 1994 27 persons died and 20 were injured.

At present, heated discussions on the transport of Kazakstan and Azerbaijan oil through the Black Sea are under way. Presently, 45,000 ships cross the straits annually, a rate of 125 per day. At the Tengiz oil fields in Kazakstan 100,000 barrels of oil per day is expected to be extracted soon and that figure is expected to go up to 700,000 in 2010 (Fig. 2). The petrol-loading facilities are expected to increase tremendously in Kuban (Russia) and Odessa-Novorosisk (Ukraine) very shortly [53]. In Kuban 40 million tons of petrol is produced [45]. A terminal with a one million metric ton capacity is being built on the Taman peninsula for the transportation of liquid gas. It will pass through the Kuban region where seismic activity of 9 on the Richter scale and earthquakes as high as 10 have been experienced [45].

The dangers lying ahead for the Black Sea are clear from these developments. The threats it poses for this semi-enclosed sea are great but economic gains always take priority over protecting the integrity of fragile ecosystems such as the Black Sea. It is difficult for the scientists in the region to determine when "enough is enough" with regards to this "business as usual" approach in the face of political pressures.

Fig. 2. Petrol pumping lineconnecting Kazakstan with Black Sea port

Pollution Sources

During the last few decades conditions in the Black Sea have changed dramatically, because of changes in the chemical composition of its waters [37, 1]. The chemical pollution load through the Danube River alone has increased 5 to 10 times [22] and it increased considerably after the Rhine-Black Sea canal connection went into full operation. The values of organic carbon and phosphorus published by Deuser [11] and Fonselius [15] are no longer collected. Human pressure on the biology and chemistry of the Black Sea has intensified as a result of organic matter production. A high level of eutrophication is expected in the coming years which will likely be followed by complex hydrochemical development changes [42] in the economically important oxygenated zone between 0-200 m [20, 1].

From 1960 onwards, an increase in petrochemical production in the Black Sea region as well as the Caspian Sea has resulted in the nomination of the former as one of the most troubled seas in the world [25]. The effluent up to 500 m^3 from the non-nuclear Chernobyl (the Kuban region) constitutes a great threat together with radionuclide wastes like RA 226 (+104 BK/kg) and Th 232 (+240 BK/kg) vis-â-vis toxic chemicals sprayed on paddy fields [45]. Thus, Kuban is an important point source of chemical pollutants, including radionuclides, reaching the Black Sea. The results published by Bodeanu [4] reveal that the input of phosphates has increased from 12.6 metric tons (1950) to 30.4 metric tons (1987) and that of nitrates from

143 x 10^3 metric tons to 741 x 10^3 metric tons during the same period along the Rumanian coast alone. The data reported by Codispoti et al. [8] show that the nitrate content in 1960 was 0.1-0.3 μg/l and in 1991 0.6-0.7 μg/l . Pollution sources of industrial origin to the Black Sea from Turkey are given in Table 1.

TABLE 1. Pollution Sources on the Turkish Black Sea Coast

Sources	# of establish-ments	Sources	# of establish-ments
Cement production and Brick Manufacturing	73	Plastic Goods	29
Food, Fruits Vegetables and Rice Processing	53	Forest. Paper and related industry	24
Hazelnut Processing	52	Rubber and Tires	14
Leather Processing	52	Milk, Fodder and Sugar Factories	11
Flour Mills	50	Sea Products	9
Copper, Iron, Steel, Metal Products and related Indus.	50	Beverages	8
Tea Processing	38	Cigarette Factories	3
Production of Fertilizers and Chemicals	30	Coal Excavation	1

An influx of water from the Black Sea into the Sea of Marmara is 371 km^3/yr and only 176 km^3/yr in the reverse direction [24]. This influx of water brings with it in an organic load of pollutants which is reported to be 1.2 x 10^4 tons per year for total phosphorus, 1.9 x 10^5 for total nitrogen and 1.5 x 10^6 for total organic carbon [33].

Ecological Diversity

Present ecological conditions and biological diversity of the Black Sea are a result of past and present hydrological changes. It is the world's biggest oxygenless water body [44] with a volume of 537,000 km^3 and an average depth of 1271 m [55]. Calcium carbonate influx from the rivers entering this sea makes it more basic (PH 7.5-8.5) as compared to that of the open sea. The temperature of surface waters changes from summer to winter in relation to latitude. Summer temperatures show a slight variation, being 21°C in the north and west and 24°C in the south and east, but in winter it drops to 2°C in the former and 12°-13°C in the latter direction. At 75 m depth the sea's temperature is constant throughout, around 8.5-9°C. Salinity values change from 19‰ in the east and middle parts to 10‰ in the north and west, where streams enter this sea. At a depth of 200 m, salinity is around 22‰. The level of water changes by 20-30 cm on a seasonal basis, (higher in early summer and lower in winter), as a result of the large input of waters from the rivers which is around 400 km^3/year with a 67% contribution from the Danube (203 km^3/year), Dnieper (54 km^3/year) and Dniester (9.3 km^3/year). The area between the Danube inlet and Odessa is the most biologically productive because of upwelling created by southwest winds up to a depth of 100 m along a 0-15 km wide zone of the coast [17, 44]. Upwelling is also observed between the Straits of Istanbul and Sinop [13]. Upwelling means great economic prospects for fisheries.

The Black Sea has a narrow live belt which is poor in taxa but rich in abundance and productivity. There are 1,200 taxa as compared to 7,000 in the Mediterranean Sea. All of the normal living organisms exist in 8% of the total water volume which covers the oxygenated part and extends down to a depth of 150-200 m. Primary productivity values are 3 mg/l as against a mean value of 1.5 mg/l given for other marine water bodies of the world. The highest productive areas are the Sea of Azov, Odessa Bay and coastal parts of the Caucasus, because of the mineral enrichment from waters of these areas. Over 350 phytoplankton species live in the shallow waters of this sea [5, 40], 70% being diatoms. The number of algal taxa is reported to be 417 (190 are red algae, 104 brown, 99 green and 20 blue-green). Along the Turkish coast, 261 algal taxa thrive (140 red, 53 brown, 50 green and 13 blue-green [16]). We have found only four typical littoral benthic communities of algae, compared to 30 in the Mediterranean.

The Black Sea has always been important for the bordering nations with regard to fish production. Of an annual total production of 1.5 million metric tons only 10% belongs to Turkey. This primarily includes 14 species out of a total of 156. In general 75% of the fish comes from the Mediterranean but sturgeon, herring and gobby are accepted as the old remnants whereas carp, sheatfish, speckled trout and tench are relatively newer species. Major representatives of fish species are atherine, shad, red mullet, needlefish, scad and anchovy. Species like bonito, mackerel, bluefish, swordfish and tunny keep migrating into and out of the sea.

Because the water has changed from brackish to saline during the changes from an open to an inland sea, mussels like Congena, Cardium, Mactra, Tapes, Trochus and Rissoa dominated under brackish water characteristics as compared to saline water organisms like Coralle, Brachiopoda, Cephalopoda and Selachie. In 1949 ships coming from Asia introduced *Rapana thomasiana* through their bilge waters. This species destroyed most of the mussel stocks. A voracious zooplankton predator - a ctenophore (*Mnemiopsis mccradyi*) - was accidently introduced in 1987 in the Black Sea from the Northwest Atlantic and is thriving [48]. Its introduction has resulted in a decrease in the biomass of copepods and zooplankton.

However, biomass of some zooplanktonic taxa have increased from 2.56 mg/m^3 to 155.56 mg/m^3, like microbial populations and phytoplanktonic species [2, 38, 39, 54]. Red tide is a common phenomenon in the sea [46]; and Chlorophyll-a has increased three times since 1970 [49] because of eutrophication.

Impact of Pollution

The Black Sea is facing a real disaster because of an increase in the organic matter, heavy metals, pesticides, phenols, oil-based products and radionuclides. There has been a great decrease in the anchovy catch and other pelagic fish in the last decade [7, 18]. The anchovy catch dropped from 295,000 metric tons (1989) to 66,000 metric tons (1990) in Turkey. In 1980 twenty-six species of fish were on the market. That number dropped to five in 1990 [25]. Oily algal taxa and sea phanerogams (flowering plants in the sea), oil-tainted fish, oily beaches and oil-covered sea birds are indicators we commonly see today along the coastal areas as a result of pollution from marine traffic activities.

Especially pernicious is the effect of transport pollution on fish resources. The dolphin is very rarely seen now. The oil pollution problem is becoming increasingly acute day by day; spawning or nursery grounds and hatcheries are adversely affected as are the recreational beaches. Generally, 10 ml/l of oil on the sea's surface can prove lethal for the spawning of turbot, belted bonito and mackerel [23]. It also decreases florescence of the sea which results in a decrease in oxygen production, which in turn causes hypoxia which again in turn leads to secondary H_2S production in addition to that which exists in the Black Sea below 150-200 m. Its effects, however, depend on the type, magnitude, time and micro-climatic conditions of the area.

Generally, aromatic hydrocarbons are poisonous for many marine species, in particular, crustaceans in deeper waters. Low concentrations can lead to a loss of sense of smell and taste by plugging the receptors in many species. Some animals become blind. An increase in the major biogenic elements like N, P, K enhances eutrophication which leads to a spontaneous spreading of microalgae. This produces "algotoxins" thereby poisoning both zooplankton and fish. The Danube alone adds 1-2 million metric tons of biogenic elements to the sea. Fish with algotoxins that are consumed by humans lead to "toxemia" which can prove lethal. Increase in NO_x around the beaches is hindering people from swimming on the Bulgarian coast [24]. Along the former Soviet Union's Black Sea coasts, the radioactive pollution of beaches discourages swimming.

Studies on the accumulation of heavy metals by marine organisms have attracted the attention of several investigators notably Bryan [6], Windom and Smith [51], Phillips [32], Portman [34], Reish and Oshida [35] and Türkan et al. [47]. FAO [14], too, has published some data related to this topic. Heavy-metal accumulation studies undertaken at some stations on the Turkish Black Sea coast reveal values below the limits set by public health authorities but they are steadily increasing. The values given on a wet-weight basis for crab *(Eriphia verrucosa)*, shad *(Alosa bulgarica)*, bass (*Serranus hepatus*) and mullet (*Mullus barbatus*) by Öztürk and Bat [28] and by Kìratli and Yìldìzdağ [19] show the following ranges:

Eriphia verrucosa (Crab)			*Alosa bulgarica* (Shad)		
Zn	9.11-10.77	µg/g	Zn	1.65-4.48	µg/g
Cu	2.42-2.72	µg/g	Cu	0.26-0.52	µg/g
Cd	0.14-0.213	µg/g	Cd	0.19-0.47	µg/g
Ni	1.10-1.61	µg/g	Ni	0.84-2.73	µg/g
Pb	0.35-0.54	µg/g	Pb	0.18-0.74	µg/g
Fe	2.33-2.89	µg/g	Fe	1.61-9.14	µg/g
Mn	0.08-0.22	µg/g	Mn	0.18-0.44	µg/g

Serranus hepatus (Bass)			*Mullus barbatus* (Mullet)		
Hg	1500-3600	µg/g	Hg	255-3912	µg/g
Cu	3-7	µg/g	Cu	1.9-10	µg/g
Zn	10-19	µg/g	Zn	2.7-10	µg/g

Öztürk and Bat [29] have reported the following results for *Ulva lactuca* (sea lettuce) on dry weight basis; 158-445, 15-127, 13-101, 14-100, 15, 84, 8-32 and 0.15-88 µg/g for Fe, Zn, Ni, Cu, Mn, and Cd respectively. Some data on pesticide accumulation has also been published covering anchovy, scad, striped goatfish, red mullet, mussel, turbot, whiting and gray mullet [30].

The State of Environment in the Caspian

The Caspian Sea, the biggest lake in the world whose waters wash the shores of Azerbaijan, Iran, Kazakstan, Russia and Turkmenistan, is at present a center of great attraction for the world's major oil companies. However, like the Black Sea, it too is on the verge of ecological crisis due to domestic, chemical, thermal and radionuclide pollution. Over 100 million m^3 of heated water flows into the sea from thermal power plants of the Apsheron peninsula [21] where 104,373,000 m^3 of sea-water and 346,000 m^3 of fresh water are used. The Bay of Baku is not much different from the bays of Izmit and Izmir in Turkey, e.g., they are equally as polluted. Reports reveal that the sea level has followed a catastrophic rise of 2.5 m since 1977, resulting in a loss of more than US$150 million in Atrav, Kazakstan alone, where thousands of hectares of cultivated land and grasslands have been inundated along with human settlements. The current rise in sea level is expected to continue till the year 2010, when it will affect 7 cities and 35 settlements with a population of 700,000 within Azerbaijan alone. It will also adversely affect off-shore oil platforms, as well as coastal railways and ports. However, the only beneficial aspect of sea level rise will be the development of flora and fauna reserves along the Caspian coast.

Pollution in this sea has increased to the extent that, like the Black Sea, out of 32 species reported in earlier fish catches, several have become extinct. The situation is worsened further by oil extraction which amounts to 12 million metric tons per year along the Azerbaijan coast alone. On an annual basis 100,000 metric tons of oil pollution run into the Caspian in addition to more than 10 billion m^3 of sewage from Armenia and Georgia [31]. Transportation of 27 million metric tons by ships from Azerbaijan depicts the situation in the transportational activities in the Caspian.

Conclusions

A sharp increase in the anthropogenic pressures on aquatic ecosystems during the past few decades has created serious problems around the world. Such ecosystems are no longer considered as inexhaustible suppliers of resources. The Aral, Caspian and Black seas are like beads in a chain of aquatic ecosystems. If steps are not taken for their conservation, our planet will face a string of major ecological catastrophes. The Black Sea basin has been a common resource for the nations surrounding it. We need to mobilize the public, governments and industries to pursue a sustainable ecological restructuring of these enclosed or semienclosed seas. For this purpose the following questions must be addressed:

a. What are the carrying capacities and resilience and vulnerability features of these ecosystems?
b. What are the possibilities for joint ecological monitoring?
c. Identify the status of present technologies used here and the possibilities for application of potentially feasible technologies for eco-reconstruction together with barriers for their application.
d. Identify ways to improve the interactions among the public, policy makers, industries and scientists.
e. Develop regional policies that foster natural world heritage so that these seas can receive protection from degradation by the international community.
f. Is it possible to group enterprises manufacturing the same type of products on a regional basis?

The steps that could be taken are as follows:

1. Environmental safety network needs to be established for the whole basin.
2. Recycling of wastes should be started with high quality technology and this could be shared on the basis of industries in different countries.
3. Common regional environmental standards should be developed.

4. Pollution mitigation programs should be started immediately with the support of the international community.
5. Coastal planning should be restructured on a regional basis to reduce the pressure of urbanization and recreation along the affected coastal areas.
6. Sustainable tourism, aquaculture and fishing should be the objective for the basin.
7. Industrial, trade, agricultural, military and other strategic activities should be integrated regionally for the welfare of regional inhabitants.
8. The number of nature reserves along the coast should be increased, particularly in Turkey, to save regional ecological diversity.
9. Management programs for coastal waters, littoral ecosystems and terrestrial ecosystems should be developed.
10. A network should be established to identify coastal risk assessment and response strategies.

References

1. Balci, A., F. Küçüksezgin, O. Altay, and A. Kontaş, 1994: Chemical changes in the pelagic ecosystem of the Black Sea. *Fen Fakültesi Dergisi*, **16**(1), 1541-1548.

2. Balkaş, T., et al., 1990: State of marine environment in the Black Sea region. Nairobi, Kenya: UNEP Regional Seas Report Study No. 124.

3. Baştürk, O., S. Tuğrul, A. Yilmaz, and C. Saydam, 1990: Health of the Turkish Straits: Chemical and environmental aspects of the sea of Marmara. ODTU-*Deniz Bilimleri Enstitüsü*, **94**(4), 9-21.

4. Bodeanu, N., 1989: Algal blooms and development of the marine phytoplankton species at the Romanian Black Sea littoral under eutrophication conditions. *Cercetari Marine*, **22**, 107-125.

5. Bolonga, A.S., 1986: Planktonic primary productivity of the Black Sea: A review. *Thalassa Jugoslavica*, **21**, 1-22.

6. Bryan, G.W., 1971: The effects of heavy metals (other than mercury) on marine and estuarine organisms. *Proceedings of the Royal Society of London*, **177**, 389-410.

7. Caddy, J.F., 1992: Update of the fishery situation in the Black Sea and revision of the conclusions of the 1990 GFCM Studies and Reviews No. 63. In: *A perspective on Recent Fishery-Related Events in the Black Sea*, J.F. Caddy and H. Griffiths. Unpublished manuscript. Rome, Italy: FAO.

8. Codispoti, L.A., et al., 1991: Chemical variability in the Black Sea: Implications of continuous vertical profiles that penetrated the oxic/anoxic interface. *Deep Sea Research*, **38**, Supplement 2, 5691-5710.

9. Demirsoy, A., 1995: Anatolian fauna IV, marine organisms. *Bilim ve Teknik Dergisi, Tübitak Yayinlari, Sayi*, **326**, S72-S77.

10. DTO, 1994: *Marine Sector Report 1993*. DTO yayin No. 36, *Türkan Matbaasi*, 340 pp.

11. Deuser, W. Ö., 1971: Organic carbon budget of the Black Sea. *Deep Sea Research*, **18**, 995-1004.

12. DIE, 1995: *Statistical Yearbook of Turkey 1994*. Publication No. 1720. State Institute of Statistics Prime Ministry, Republic of Turkey, DIE Matbaasi, 752 pp.

13. Durukanoğlu, H.F., C. Erüz, 1994: Wind-induced upwelling of southeast (Trabzon) coast of Black Sea. E.Ü. *Fen Fakültesi Dergisi*, **16**(1), S1569-S1577.

14. FAO (Food and Agriculture Organization), 1983: *Compilation of Legal Limits for Hazardous Substances in Fish and Fishery Products*. Circular No. 764. Rome, Italy: FAO.

15. Fonselius, S.H., 1974: Phosphorus in the Black Sea. In: *The Black Sea Geology, Chemistry, and Biology*, E.T. Degans and D.A. Ross (eds.). Vol. 20, 144-150. American Association of Petrol and Geology.

16. Güner, H., and V. Aysel, 1996: *Marine Benthic Vegetation: Turkey. Ecological Studies*. Vol. 123. Heidelberg, Germany: Springer-Verlag.

17. Hedgpeth, J.W., 1957: Treatise on marine ecology and paleoecology. *The Geological Society of America*, **1**, 801-899.

18. Kideyş, A.E., 1994: The devastated ecosystem of the Black Sea and its impacts on anchovy fisheries. *E.Ü. Fen Fakültesi Dergisi*, **16**(1), S993-S1002.

19. Kiratli, N., and H.G. Yildizdağ, 1994: Hg, Zn, and CU contents in the tissues of *Serranus hepatus* Lin and *Mullus barbatus* Lin. *E.Ü. Fen Fakültesi Dergisi*, **16**(1), 159-171.

20. Kirikova, M.V., 1986: The content of inorganic nitrogen forms in the upper layer of the Black Sea in the later summer period. *Marine Ecology*, **23**, 3-10.

21. Mansurov, A.E., 1993: *State Report on Environmental Condition and Nature Protecting Activity in Azerbaijan Republic*. Baku, Azerbaijan: State Committee of the Azerbaijan Republic on Nature Protection.

22. Mee, L.D., 1992: The Black Sea in crisis: The need for concerted international action. *Ambio*, **21**(4), 278-286.

23. Mielke, J.E., 1990: *Oil in the Ocean: The Short- and Long-Term Impacts of a Spill*. Washington, DC: Congressional Research Service Report for Congress.

24. Nedialkov, S., 1994: Ecological problems of the Black Sea. *Proceedings, Central Asia and Black Sea Environment Conference, 1993*. Japan: Sasakawa Foundation, S55-S58.

25. Nogaideli, Z.T., 1994: Environmental situation in Georgia. *Proceedings, Central Asia and Black Sea Environment Conference, 1993*. Japan: Sasakawa Foundation, S67-S72.

26. NRC (National Research Council), 1985: *Oil in the Sea: Inputs, Fates and Effects*. Washington, DC: National Academy Press.

27. Nurlu, E., Ü. Erdem, F. Özdemir, M. Öztürk, and H. Güner, 1996: *Impacts of Sea Traffic on Turkish Littoral Ecoysystems*. Montpellier, France: Okeanos.

28. Öztürk, M., and L. Bat, 1994 a: A study on the heavy metal levels in ulva lactuca (L.), Le jolis 1863, collected from Sinop coast of Black Sea. E.Ü. *Fen Fakültesi Dergisi*, **16**(1), S187-S195.

29. Öztürk, M., and L. Bat, 1994 b: The trace element levels in some edible organisms in Sinop coasts of Black Sea, Turkey. E.Ü. *Fen Fakültesi Dergisi*, **16**(1), S177-S186.

30. Öztürk, M., and I. Türkan, 1989: *The Role of Biocides in Turkish Agriculture Proceedings, Plants and Pollutants in Developed and Developing Countries.* Bornova-Izmir: Ege University Press, p.7.

31. Panahov, S., 1994: Ecological problems and sustainable development situation in Azerbaijan. *Proceedings, Central Asia and Black Sea Environment Conference, 1993.* Japan: Sasakawa Foundation, S48-S52.

32. Philips, D.J.H., 1976: The common mussel *M. edulis* as an indicator of pollution by zinc, cadmium, lead, and copper. I: Effects of environmental variables on uptake of metals. *Marine Biology*, **38**, 59-69.

33. Polat, Ç, and S. Tuğrul, 1994: A comparison of Black Sea and other sources as inputs of total phosphorus, nitrogen, and organic carbon to the Sea of Marmara. E.Ü. *Fen Fakültesi Dergisi*, **16**(1), S273-S282.

34. Portman, J.E., 1976: *The Role of Biological Accumulators in MonitoringProgrammes, Manual of Methods in Aquatic Environment Research: Part 2: Guidelines for Use of Biological Accumulators in Marine Pollution Monitoring.* FAO Fisheries Technical Report No. 150, 1-7. Rome, Italy: FAO.

35. Reish, D.J., and P.S. Oshida, 1986: *Manual of Methods in Aquatic Environment Research.* FAO Fisheries Technical Report No. 247. Rome, Italy: FAO.

36. Sapozhnikov, V.V., 1990: Ammonia in the Black Sea. *Oceanology*, **30**, 39-42.

37. Serez, M., 1983: *The Black Sea Coast and Pollution Problems.* Coastal Management Seminar, 19-41. Karadeniz Technical University Press, Trabzon, Turkey.

38. Sushkina, E.A., and E.J. Musayeva, 1990a: Structure of plankton community of the Black Sea: Epipelagic zone and its variation caused by invastion of a new *Ctenophore* species. *Oceanology*, **30**(2), 225-228.

39. Sushkina, E.A., and E.J. Musayeva, 1990b: Increasing abundance of the immigrant *Ctenophore mnemiopsis* in the Black Sea: Report of an expedition by the R/Vs Akvanavt Hidrobilog in April 1990. *Oceanology*, **30**(4), 521-522.

40. Sushkina, E.A., and M.Gl Vinogradov, 1991: Plankton changes in the open Black Sea for many years. *Okeanologia*, **31**, 973-979.

41. Sorokin, M.I., 1983: The Black Sea. In: *Estuaries and Enclosed Seas: Ecosystem of the World*, B.H. Ketchum (ed.), 253-291. Amsterdam: Elsevier.

42. Tuğrul, S., Ö. Baştürk, C. Saydam, and A. Yilmaz, 1992: Possible changes in the hydrochemistry of the Black Sea inferred from density profiles. *Nature*, **359**, 137-139.

43. Tuğrul, S., 1983: *Environmental Problems of Turkey.* Ankara: Türkiye Çevre Sorunlari Vakfi Yayini. Önder Matbaa. 336 pp.

44. Tolmazin, D., 1985: Changing coastal oceanography of the Black Sea. 1. Northwestern Shelf. *Prog. Oceanography*, **15**, 217-276.

45. Tsipshenko, N.M., 1993: Environmental problems of Kuban area. *Proceedings, Central Asia and Black Sea Environment Conference, 1993.* Japan: Sasakawa Foundation, S99-S100.

46. Tumantseva, N.I., 1985: Red tide in the Black Sea. *Oceanology,* **25**(1), 99-101.

47. Türkan, I., M. Öztürk, and A. Sukatar, 1989: Heavy metal accumulation by algae in the Bay of Izmir, Turkey. *Revue, Inernationale d'Oceanographie Medicale Tomes* , LXXXiii-LXXXiv, 71-76.

48. Vinogradov, M.Y.E., et al., 1989: A newly acclimated species in the Black Sea: The *Ctenophore Mnemiopsis eildyi (Ctenophora: Lobata). Oceanology,* **29**(2), 220-224.

49. Vinogradov, M.Y.E., 1990: Investigation of the pelagic ecosystem of the Black Sea. 44th Cruise of the R/V Dmitriy Mendeleyev, 4 July-17 September 1989. *Oceanology,* **30**(2), 254-256.

50. Vinogradov, M.Y.E., E.O. Arashkevich, and S.V. Ilchenko, 1992: The ecology of the *Calanus ponticus* population in the deeper layer of its concentration in the Black Sea. *Journal of Plankton Research,* **14**(3), 447-458.

51. Windom, H.L., and R.O. Smith, 1972: Distribution of iron, magnesium, copper, zinc, and silver inoysters along the Georgia coast. *Journal of Fisheries Research Board of Canada,* **29**(4), 450-452.

52. Yamaguchi, H.D., 1990: *The Changing Environment for Seaborne Oil Trade: An Overview of Flow Developments nad Forecast Tonnage Requirements through the 1990s.* Honolulu, Hawaii: East-West Center Energy Program. 48 pp.

53. Zabalin, S., 1994: Socio-ecological association: Central Asia and Black Sea. *Proceedings, Central Asia and Black Sea Environment Conference, 1993.* Japan: Sasakawa Foundation, S95-S98.

54. Zaika, V.E., and M.G. Sergeeva, 1990: Morphology and development of *Ctenophore-Colonizer Mnemiopsis mccradyi* (*Ctenophora, Lobata*) in the Black Sea. *Zool. Zh.,* **69**(2), 5-11. (In Russian)

55. Zenkevitch, L., 1963: *Biology of the Seas in the USSR.* London, UK: George Allen and Unwin. 955 pp.

ENVIRONMENTAL PROBLEMS OF THE BALTIC SEA REGION

SERGEI PREIS
Department of Chemical Engineering
Tallinn Technical University
Tallinn, Estonia

Abstract

The Baltic Sea and the Caspian Sea have very much in common. Firstly, both are shallow seas with brackish water and restricted water exchange, which makes them more sensitive to pollution. Secondly, the drainage areas of both seas are densely populated and heavily industrialized. Thus, the present process and rates of pollution constitute a threat to the people, flora and fauna of all the countries around these seas. Thirdly, the Baltic Sea and the Caspian Sea are also very productive seas with intensive fishing. Finally, both regions constitute fine examples of eco-geographical units. Thus, if we aspire to improve the environmental conditions of the Baltic and the Caspian, we have to do this in a coordinated manner that includes all countries in each of the drainage basins.

Physical Geography of the Baltic [5]

The Baltic Sea is a brackish, non-tidal, epicontinental sea in a former glaciated area. Epicontinental means that the sea is actually situated on the continent rather than between continents. Others of this type are the Hudson Bay, the Persian Gulf and the North Sea. These are all shallow seas, rarely more than 100 m deep, whereas the seas between continents are generally at least 2,000 m deep (e.g., the Mediterranean, the Black Sea and the Red Sea). An epicontinental sea is more or less cut off from the oceans.

Brackish waters are rather unusual on the earth and only a relative few animals and plants are adapted to live in them. A strong tidal range in the Kattegat would have increased the salt influx to the Baltic, but the tide is only about 1-2 dm, because a tidal node preventing the tidal waves from entering the Kattegat is situated between Norway and Denmark.

M.H. Glantz and I.S. Zonn (eds.),
Scientific, Environmental, and Political Issues in the Circum-Caspian Region, 227-244.
© 1997 *Kluwer Academic Publishers. Printed in the Netherlands.*

The latest glaciation reshaped the landscape and pressed down the crust of the earth. Once the ice disappeared, the land began to rise and continues to do so. New bottom areas are then exposed to waves. It means that nutrients bound to these old sediments may, once again, be used by plankton and algae. Hence, the few species living in the Baltic have an ample supply of food and they are generally present in large numbers. Fishing in the Baltic is intense and of considerable economic importance.

The Baltic is located at the high latitudes and one characteristic feature of the Baltic is its ice. Another unique feature is the archipelagos — the archipelago area outside Stockholm (Sweden) consists of more than 25,000 islands. This means, among other things, that there is a great potential for sailing, fishing and recreation.

The drainage basin is densely populated (80 million people) and heavily industrialized. The region between Berlin and Krakow drains to the river Odra and the upper reaches of the Wisla entering the Baltic. The measures against environmental damage in this region, as well as in most parts of Estonia, Latvia, Lithuania and Russia leaves very much to be desired. The Baltic Sea is heavily polluted, and also sensitive to pollutants because of, for example, the cold climate and its restricted water exchange with the ocean.

So, there are many reasons why the Baltic is such a unique aquatic ecosystem: (1) it is a large, (2) very shallow and (3) sheltered inland sea with (4) brackish water, (5) many coastal types, (6) in a cold climate; (7) the catchment area is heavily industrialized, with a large population and intensive land use; (8) the Baltic Sea is also a most productive sea with intensive fishing; it is (9) sensitive but with (10) great recreation potential; and (11) the present pollution constitutes a threat to the people, flora and fauna of all the littoral countries.

The Baltic ecosystem is very dynamic and has gone through comprehensive modifications. This is partly a result of changes in agriculture, forestry, public transport, and industry and in a number of other activities. But the Baltic has gone through considerable natural changes which largely involve land uplift — new islands and even entire coastal and archipelago areas have appeared as a result of the fact that the ground today is "rising" after having been pressed down and "planed-off" by the massive glacial inland ice. Land uplift and the quaternary glaciation are still of major importance for the Baltic ecosystem.

The Baltic region constitutes a fine example of an eco-geographical unit. Thus, if we aspire to improve the environmental conditions of the Baltic Sea, we have to do this in a coordinated manner that includes all countries in the drainage basin.

An important parameter — oxygen content — deserves a few comments. The surface water is generally well oxygenated with O_2-concentrations between 7 and 9 mg/l; the values for January are often higher than for September, depending on the high production of organic material during the summer months. When the algae and plankton die, they are consumed by bacteria and in this decomposition/mineralization of dead organic matter oxygen is utilized. This results in lower O_2-values in September. From about 60-70 m water depth, we may note a very steep O_2-gradient; and at a water depth of about 140 m we can see that there is no oxygen at all left; that is the so-called redoxcline. Beneath the redoxcline, hydrogen sulphide is produced in biochemical processes.

Oxygen Deficiency Is Killing Cod

The cod population in the Baltic has decreased significantly since the early 1980s. Why? It certainly depends on many factors in a complicated system of interactions, but one of those factors is linked to the oxygen conditions.

Each female cod lays many eggs. In the salty water of the Kattegat, the roe remains suspended in the surface, oxygen-rich, water. In the brackish surface waters of the Baltic the salinity and, hence, also the density, is not high enough to allow the roe to float. It sinks to the saline bottom water until this descent is stopped by the greater density of the water. If this bottom water is oxygen-rich, there will be a good spawning year for cod, but if the oxygen content is low, the eggs will die and the spawning will fail.

Not since 1977 have events of salt water inflows managed to improve the oxygen situation in the deep areas in the Baltic Proper for any long period. The production of organic material, algae, plankton, bacteria, animals, has been so great that the addition of oxygen that still enters from the Kattegat is quickly utilized in the degradation of this continuously increasing "biological rain" down in the deep areas.

Land Uplift - Still a Key Process Today

Thousand-year-old sediments influence the Baltic ecosystem today. When the old bottom rises after having been depressed by glacial ice, it will eventually reach the so-called wave-base, which is the water depth at which waves can exert a direct influence on and re-suspend the sediments. The wave-base depends on the duration and velocity of the wind; during storms it may reach water depths of 50-60 m in the Baltic Proper. Although land uplift in the Baltic Proper is only on average 1.5 mm per year, this implies that large amounts of old glacial and post-glacial sediments are eroded. In this way, the carbon, nitrogen and phosphorus contained by these sediments will again enter the ecosystem of the Baltic. Recent measurements demonstrate that as much as 80% of the material settling onto the deep bottoms may be old eroded material. The new supply of material to the sea from the rivers and direct discharges and from bioproduction in the system contributes only about 20% of the amount annually accumulating on the bottoms in the deep areas.

Life in the Baltic Sea

THE SPECIAL CONDITIONS OF THE BALTIC [6]

The Baltic Sea is one of the largest brackish water (salinity around 0.5 to 0.6%) areas in the world. It is also a relatively young sea, created after the Ice Age, with the present low-salinity conditions largely unaltered for only some 3,000 years. In evolutionary terms this is a very short time.

As compared to fully marine environments, the Baltic Sea with its brackish water is very poor in both flora and fauna. The number of marine species found decreases dramatically as one goes through the Danish Straits into the Baltic Proper and then continues to decrease as we continue up to the Finnish Gulf, through the Archipelago Sea into the Bothnian Sea and, finally, the Bothnian Bay. Of the 1,500 macroscopic animals off the Norwegian coast, only about 70 species are still to be found in the central Baltic. Of the approximately 150 macroscopic algae on the Norwegian coast only 24 remain at the Finnish coast.

Since the number of species in the Baltic is low compared to fully marine systems, the number of "stand-ins" capable of filling the niche of an excluded species is small or even zero. The risk that whole life forms may be excluded, because of environmental disturbances, is therefore great. The continuously decreasing salt concentrations, the salinity gradient, is the main reason for this poverty of life forms. Only very few of the marine organisms, adapted to a salt content of 3.5%, are able to regulate the water content in their cells when the osmotic pressure is changing due to the diminishing levels of salt.

Temperature also influences life in the Baltic. Thus, the salinity gradient is paralleled by a productive season of only 4-5 months in the far north of the Bothnian Bay caused by up to six months of ice cover, and a productive season in the southern sounds and Kattegat of 8-9 months with sea ice only in cold years. Thus, in some instances temperature rather than salinity may be the factor limiting any further penetration of marine organisms into the Baltic.

FISH AND FISHING

The fish fauna of the Baltic Sea exhibits a mixture of species of freshwater and marine origin, similar to the rest of the flora and fauna. Again, the salinity decrease causes the Baltic to be poor in species. The number of marine fish species in the North Sea is 120, in the Kiel Bay 69, in the southern and middle parts of the Baltic Sea 41, in the Gulf of Finland and Gulf of Bothnia, 20.

The fish in the open sea are dominated by three marine species — herring, sprat and cod. Herring migrates during the year to different parts of the Baltic ecosystem to search for food and reproduction areas. Winter time is dominated by decomposition processes. Due to a lack of food, herring is forced to feed at the bottom, successively moving to shallow areas where they stay until spawning in May. The *sprat* has a life cycle very similar to that of herring. However, it spawns in the open sea and has pelagic roe like the cod, although sprat roe stays in shallower water with lower salinity. Compared to herring, sprat is a more specialized zooplankton feeder. There was a drastic increase in the catches of sprat from the mid-1950s to mid-1970s, most probably due to eutrophication. Since this period, catches have decreased significantly, probably due to a combination of heavy predation by cod and overfishing. The reproduction of cod in the Baltic is restricted to areas with salinity around 1‰ and higher, which are found mainly in five deep basins. The resulting increased primary production, causing oxygen depletion in deep water, decreased the reproductive success of cod.

MAMMALS: THREATENED SPECIES

Predators other than fish are air-breathing vertebrates such as sea birds, seals and whales. In the Baltic three species of seal are living: the grey seal, the ringed seal and the harbour seal. The grey seal is the largest and the ringed seal is the smallest species. Seal populations have been greatly reduced in recent years. The number of ringed seal is estimated at about 6,000. The ringed seal is mainly found in the Bothnian Bay, Bothnian Sea, Gulf of Finland and Riga Bight. The ringed seal is closely related to the Arctic populations and is considered a glacial relict. The number of Baltic grey seal living in the archipelagos along the Baltic coast were around 2,000 animals in 1990. The harbour seal inhabits only the southernmost part of the Baltic Sea where the estimated number is 200 individuals. During the spring of 1988, many of the harbour seals along the coast of Kattegat and Skagerak were killed by a virus. The only whale previously common, the harbour porpoise (*Phocoena phocoena*), is now virtually extinct in the Baltic.

BIRDS

The Baltic region is rich in bird species. Some 340 species are found regularly in the region. Many of them are water fowls living in the Baltic. Others, such as waders, live along the coast or in surrounding wetlands. Compared to other marine environments, the Baltic is rich in species, with its combination of marine and freshwater birds. This is a sharp contrast to the underwater animal communities.

POLLUTION SENSITIVITY

The assumption that the species of the Baltic Sea are more sensitive to stress by pollution than organisms in the North Sea has been suggested several times. In the few experiments that have been performed this has also been the case. So far, we do not know of any instance where North Sea individuals have been more sensitive to environmental stress than the Baltic Sea individuals of the same species.

There are several explanations for the increased sensitivity of the Baltic species. In summary, one might say that the pollutants interfere with a system that is already under stress from its adaptation to lower salinity. An example is the interactions of toxic substances with membrane biomolecules. These membranes are already key components for keeping the salt balance and the cell can not afford to lose their functions.

It is well known that the Baltic Sea has a low salinity and it has obviously been taken for granted that the Baltic Sea environment, like other estuaries, is also more variable than the North Sea. Although there is a clear salinity gradient from the North Sea into the inner parts of the Baltic Sea, this gradient is very stable with the main reduction in species numbers occurring outside the Danish sounds. Also, if annual variations in salinity and temperature at different depths are compared between the northern Baltic Proper and the North Sea coast, it is clear that salinity, the factor having the largest biological impact, varies more in the North Sea than in the Baltic, both annually and vertically in the upper-water layer. Temperature variations are about the same, but because of a lower freezing point in highly saline water, biologically detrimental low-water temperatures are more likely to be reached in the North Sea. The North Sea should, thus, be regarded as more variable in environmental factors, which would be the reason for the higher genetic diversity observed in populations of the North Sea as compared to those of the Baltic. In all cases increased genetic diversity in a species was correlated with higher stress tolerance.

The low number of species together with little genetic variation in many of the species occurring in the Baltic make the sea more sensitive to pollution. Since both animals and plants live under salinity and partly under temperature stress, some of them almost at their outer limits of survivability, additional stress by different polluting substances might be even more fatal in the Baltic than in the environments that are more marine-like such as the Kattegat and North Sea.

Eutrophication of the Baltic Sea [4]

Eutrophication: increased input to a water body of mainly nitrogen and phosphorus, which causes increased primary production/growth of algae and higher plants

Eutrophication causes great changes in aquatic ecosystems and undesirable degradation of water quality leading, for example, to oxygen deficit and fish kill. Eutrophication is a common phenomenon in freshwater ecosystems. It is a natural process in which nutrient poor (oligotrophic) lakes are slowly transformed into nutrient-rich (eutrophic) lakes, a process which normally could take thousands of years. Due to human activities, however, eutrophication has been accelerated ("manmade eutrophication"), and is now a world-wide problem, including semi-enclosed seas, e.g., the Baltic Sea.

During the 1960s, it became obvious that man-made eutrophication was causing an increasing and undesirable degradation of water quality in lakes and reservoirs. The growing problem of toxic blue-green algal blooms has been of special concern. Interfering with beneficial uses of waters, this deterioration has also caused significant economic losses, which motivated the Organisation for Economic Cooperation and Development (OECD) to support studies for increasing knowledge of how to control eutrophication. Since the mid-1970s, noxious and sometimes toxic algal blooms, anoxic conditions and fish kills have been increasingly reported in marine areas.

Eutrophication may alter the recreational value of surface waters and impair activities such as fishing, swimming, etc., resulting in both social impacts and economic losses. If marine eutrophication cannot be stopped, it may severely damage the production of fish and shellfish with undesirable consequences for society.

Main Problems Associated with Eutrophication

A. Water quality impairment (freshwater)

- taste, odor, color; filtration, flocculation, sedimentation and other treatment difficulties;
- hypolimnetic oxygen depletion: pH changes, increased concentrations of Fe, Mn, CH4, H2S;
- toxicity.

B. Recreational impairment (fresh and marine waters)

- unaesthetic water, turbidity;
- hazard to swimmers;
- increased health hazards.

C. Fisheries impairment (fresh and marine waters)

- fish mortality;
- undesirable fish stocks.

Nitrogen and Phosphorus: Key Nutrients in Eutrophication

Nitrogen is recycled from water to the atmosphere by a process called denitrification. No similar return process for phosphorus exists. This element "travels" through ecosystems in one way only, namely from soil to surface waters and ultimately to the sea. Man-made inputs of these nutrients have increased dramatically during this century, due mainly to three processes: (1) use of artificial fertilizers; (2) use of synthetic detergents; and (3) combustion of fossil fuels.

A considerable amount of phosphorus may become "fixed" in the sediments, often bound to iron, where it is effectively outside the biological cycle. However, depending on biochemical processes, great amounts of phosphorus may be released from the sediments. This may occur under anoxic conditions in deep water and on shallow eutrophicated bottoms at high summer temperatures. This release of phosphorus is often called "internal loading" and is a result of long-term nutrient input. Internal loading supports eutrophication and can cause high primary production even after a substantial reduction in external loading.

The external loading of nitrogen and phosphorus to all sub-areas of the Baltic Sea generally has a ratio greater than 16:1 (optimum ratio for plant growth). The ratio between these elements in winter surface waters is variable. In the Bothnian Bay the ratios are higher than those of the input loads, while in the Baltic Proper and the Kattegat the ratios are relatively lower. These differences may reflect the importance of internal processes which can regulate nutrient concentrations, for example, through a generation of nutrients from organic matter in surface waters, or benthic denitrification in shallow waters or in nearly anoxic deep waters, or low solubility and precipitation causing sedimentation.

Nutrient enrichment tests generally indicate nitrogen limitation in the Baltic Proper and the Kattegat, although algal growth was stimulated by phosphorus addition in the Baltic Proper during summer blooms of blue-green algae. Nitrogen may also be the most limiting nutrient in coastal areas of the Baltic Proper which are not directly influenced by nutrient inputs. Large-scale denitrification in the sediments is the explanation for low nitrogen to phosphorus ratios in the Baltic Proper.

In the Bothnian Bay phosphorus plays the key role as a limiting nutrient. High ratios of nitrogen to phosphorus may be explained by couplings of phosphate to iron in forest river waters discharging into this bay.

NUTRIENT SOURCES

1. Land point sources (discharge pipes from sewage treatment plants and factories)

2. Land non-point or diffuse sources:

   - runoff from urban areas
   - cropped lands (fertilizers, erosion)
   - pastures (feces, erosion)
   - manure stores
   - dairies
   - sewage (human population)
   - forests (in recent years due to high atmospheric nitrogen deposition, fertilization, deforestation, construction of forest roads and ditching)

3. Atmospheric deposition (nitrogen oxides):

- combustion of fossil fuels (oil, coal, oil-shale) in factories and power stations and for transportation (vehicles, aircraft, ships);
- combustion of biomass (to produce heat and electricity)
- evaporation of ammonia from manure on farms

4. Release from sediments (phosphorus may be released under anoxic conditions from sediments to the water, where it can contribute to algal growth).

How Eutrophication Affects the Baltic Sea

A general overview of the changes with a main focus on the coastal zone can be summarized as follows:

- increased primary production
- increased algal blooms
- increased chlorophyll concentrations
- increased deposition of organic matter on the bottom
- increased macrobenthic biomass above the halocline
- increased frequency and severity of oxygen deficiency in bottom waters
- decreased water transparency
- decreased macrobenthic biomass below the halocline

An idea of the magnitude of the changes in the Baltic ecosystem can be obtained from estimated changes in major energy flows expressed as organic carbon. The following data illustrate human impacts in the 20th century:

- increased pelagic primary production by 30 to 70%
- increased zooplankton production by 25%
- increased sedimentation of organic carbon by 70 to 190%
- an approximate doubling of the macrobenthic production above the halocline
- more than a tenfold increase in fish catches (only partly due to increased fish production, and because of increasing fishing effort)
- anoxic bottom water, which has wiped out the macrobenthos over nearly 100,000 km^2 of the deeper bottoms of the Baltic Proper and the Gulf of Finland

The most striking aspect of eutrophication may be the phytoplankton blooms, when mass development of microscopic algae over large water areas drastically reduces transparency and sometimes creates surface scums and odors. The scum-forming algae are blue-greens which easily reach the water surface.

In the summer of 1991 there was a very extensive bloom in the open Baltic and along the southern and southeastern coasts of Sweden. The bloom-forming blue-green algae may be toxic to animals. *Nodularia* produces a peptide called hepatoxin, which can degenerate liver cells, promote tumors and cause death from hepatic haemorrhage. Death of dogs caused by *Nodularia* blooms has been reported in Denmark, Gotland and the southeastern coast of Sweden. In other areas toxic algae have affected horses, cows, sheep, pigs, birds and fish. Human problems such as stomach complaints, headaches, eczema and eye inflammation have been coupled to *Nodularia* blooms.

Evidently, the nutrient conditions are excellent for many bloom-forming species. It is reasonable to predict that considerable blooms will continue to develop in the future, many of them of toxic character.

Can Eutrophication of the Baltic Sea Be Reduced or Stopped?

To reduce or stop eutrophication in the Baltic Sea is theoretically very simple. It is a question of reducing the input of nutrients into sea water. In practice, however, this is huge and very difficult task to perform.

As shown above, both nitrogen and phosphorus may play key roles in limiting algal growth in the Baltic Sea. In the Bothnian Bay in some coastal areas and for the nitrogen fixing blue-green algae during summer, phosphorus seems to be the key element. In the remaining parts of the Baltic Sea, including the Kattegat, and for the other important algal groups, nitrogen is the most limiting nutrient. To improve conditions, nitrogen loading must be reduced to lower the general production level, and phosphorus loading must be reduced to limit the growth of blue-green algae.

By How Much Must the Loading Be Reduced to Improve the Baltic Sea?

If the aim is to restore the sea to the conditions which existed before the bottom ecosystems became severely damaged, the loads must be reduced to levels prevailing at the end of 1940s! This means at least a 50% reduction of nitrogen and of phosphorus. To attain conditions in the Kattegat corresponding to those observed during the 1960s before the changes in the macroalgae communities appeared, nitrogen and phosphorus inputs must be reduced by at least 50% and perhaps up to 70-80%.

Is it Necessary to Reduce Both Nitrogen and the Phosphorus Loading?

It is known that nitrogen concentrations in the Baltic have increased more rapidly than those of phosphorus. If this trend continues, phosphorus may gradually become the key limiting element not only for growth of blue-greens but of other algae as well. Considering eutrophication in the open Baltic Sea only, and in a long-term perspective, a reduction may be achieved by decreasing only the load of phosphorus. This measure would be the simplest and most economical one but will not improve the Kattegat. In a more total environmental perspective, however, there is an absolute need also to reduce nitrogen loading from the atmosphere, agriculture and sewage. As is well known, nitrogen is causing more problems than eutrophication, as it contributes to global warming, acidification and pollution of groundwater.

WHEN CAN IMPROVEMENTS BE EXPECTED?

Nutrients in marine ecosystems may have a comparatively long turnover time. This means that there may be a delay between inputs and eutrophication effects and also between a reduction of inputs and a subsequent algal response. For the Baltic Sea it has been calculated that the phosphorus concentration will continue to increase for decades, even if inputs remains at the present level.

Political agreements within the Helsinki Commission, the North Sea Conferences and the Paris Commissions on a 50% input reduction of nutrients may result in visible improvements in coastal areas after only a few years. Regarding the entire marine ecosystem, it may take 5 to 15 years, until recovery can be noted. But a 50% reduction may not be enough to eliminate the negative effects.

Concerning the time schedule, one must also remember that it will take time to finance and build sewage treatment plants for nutrient removal, especially in East European states and newly independent republics of the former Soviet Union. Some large populations lack sewage treatment altogether, while in other areas most people are connected to phosphorus removal plants. Effective and large-scale nitrogen removal in sewage will be more difficult and more expensive to develop, which means that a rapid decrease in nitrogen loading will be very difficult to achieve. Contributing to this more pessimistic view is the fact that about 50% of the total nitrogen load on the Baltic Sea has an atmospheric origin, an input very difficult to reduce rapidly. It seems, therefore, realistic to expect a longer recovery time than that predicted above.

PROCESSES WHICH MAY IMPEDE IMPROVEMENTS:

- nutrient recycling (release from sediments)
- global warming
- the open reflux of nutrients (from land to water: fertilizers, combustion of fossil fuels, deposits of phosphorus in sludge, etc.)

Taking into consideration the above factors influencing eutrophication of the Baltic Sea, the following conclusions can be made:

- it will take a long time to restore the Baltic Sea
- the open reflux of nutrients must be substituted by more closed systems

- it is questionable if our current society and modern techniques can accomplish the tasks at hand. Only a new human lifestyle with effective management of natural resources can save the Baltic Sea, as well as all other areas plagued with large-scale environmental problems

Industrial Emissions and Toxic Pollutants in the Baltic Region [2]

The structure of industry is rather different in different regions around the Baltic Sea. The development of industry has been governed by the availability of natural raw materials (water power in the North; forests and metal ores in Sweden and Finland; oil shale in Estonia and Russia; metal, salt and coal deposits in Poland). In Sweden and Finland the metal and pulp and paper industries constitute the most important branches; Denmark is dominated by the food industry; Germany has a very diversified industrial structure. Industries in all of these countries have generally advanced, modern technologies and emissions have been significantly decreased over the last 10-15 years. Because of extremely large and diverse consumption, however, the problems associated with the diffuse impact of industrial production remain to be solved. On the other hand, in Russia, Estonia, Latvia, Lithuania and Poland many industrial plants are technologically outdated and need to be rebuilt to meet the present environmental demands. Although the consumption per person is relatively low in those countries, problems also exist with waste handling. Excessive amount of nutrients and industrial pollutants are transported to the Baltic Sea by way of rivers, since the wastewaters from cities and villages as well as from industrial plants are poorly purified or not purified at all.

The quantitative load, as well as the type of pollutants, are thus considerably different in the various sub-areas of the Baltic. The load of airborne pollutants is higher in the southern part than in the northern part, due to the denser population and the higher degree of industrialization in the south, and due to more extended air transport from remote areas. There are fairly accurate data on some emissions from the major point sources in the Baltic region, with the exception of some industrial operations in the eastern countries. However, data on individual chemicals are insufficient. National statistics cover just a limited number of chemicals and a great deal of information is kept confidential. The most difficult problem is still to be found in the huge emissions from diffuse non-point sources. These are extremely difficult to monitor and are, therefore, largely unknown.

Our knowledge of the distribution and effects of hazardous substances is far from complete. Even less is known about the interactions between hazardous substances and elevated nutrient levels, the decreased oxygen content in deep water layers, and the combined effects of two or more chemicals. In addition, the Baltic Sea is a very special recipient (because of its seasonality, brackish-water, special organisms) which implies that available knowledge of effects of toxic substances from fully marine or freshwater conditions should not necessarily be applied to the Baltic ecosystem.

MAJOR SOURCES OF INDUSTRIAL POLLUTANTS

The pulp and paper industry and the metal industry are the most important sources of anthropogenic pollution in the northern part of Baltic. Ten of the fifteen Swedish pulp mills are situated along the Baltic Sea coast and another two mills are located on inland waters, which are connected to the Baltic Sea. Most of the Finnish mills are located on inland waters connected to the Gulf of Finland. The copper smelter in Rnnskr (outside Skellefte), the titanium dioxide production plant in Pori and the metal smelter in Harjavalta (outside Pori) are the most important point sources of heavy metals and inorganic acids in the Bothnian Bay / Bothnian Sea area. The Baltic Proper is heavily polluted by sources along the eastern and southeastern coast. Parts of this region are heavily industrialized and densely populated. Especially the Polish rivers, the St. Petersburg area and the industrialized centers in northern Estonia (Narva, Sillame, Kohtla-Jrve) contribute considerably to the total emissions of pollutants to the Baltic Proper.

A major portion of the pollution from Estonia, Latvia and Lithuania originates from the cities of Tallinn, Riga, Kaunas and Vilnius and is the result of poor sewage treatment. Several beaches have been closed for swimming and bathing in these countries due to extensive bacterial pollution.

The city of St Petersburg with its five million inhabitants is a major source of pollution. The total emission of chemicals into the atmosphere has been estimated to be as high as 470,000 tonnes per year, i.e., more than 1,000 metric tons daily. Approximetely 60% of the emissions originate from traffic and about 32% from various industrial activities. The wastewater discharges from the city constitute another serious problem. Approx. 4.4 million m^3 of wastewater, containing some 2,000 metric tons of chemicals, are discharged every day, i.e., 1.6 km^3 /year. The majority of the total input (73%) is domestic waste and some 50-70% of the pollutants are estimated to be discharged into the Neva River or the Neva Bay without any purification.

The intensive utilization of oil shale in northeastern Estonia is a major source of heavy metals and phenols in this area. Approximately 200,000 metric tons of fly ash, containing huge amounts of heavy metals, are emitted into the atmosphere each year. The most abundant metals in the flue gases are lead (36 metric tons/yr), zinc (36 metric tons/yr), copper (6.6 metric tons/yr), cadmium (0.8 metric tons/yr) and mercury (0.04 metric tons/yr).

Chlorinated Compounds from the Pulp and Paper Industry

The pulp and paper industry constitutes a major source of emissions of chlorinated organic pollutants to the Baltic Sea. Both Finland and Sweden are among the world's largest producers of pulp and paper.

Around 15,000 metric tons of halogenated material were discharged from pulp mills into the Baltic Sea some years ago. Other coastal industries and municipal treatment plants contribute 900-1,000 metric tons/yr. The highest concentrations of AOX (absorbable organically bound chlorine) have been found in the Bothnian Bay, while the concentration decrease moving southwards to the Baltic Proper. Process changes in pulp mills have resulted in a marked decrease in the emissions of halogenated organic matter in recent years. Chlorine has to a great extent been substituted by chlorine dioxide, a bleaching agent which produces substantially lower amounts of chlorinated by-products. Pre-bleaching with oxygen is utilized in order to reduce the amount of chlorine used later in the process. Aerated lagoons, ultrafiltration and activated sludge plants are examples of treatment stages which are increasingly used in order to reduce the concentrations of chlorinated organics in the wastewaters leaving pulp mills. The above-mentioned measures have resulted in a decrease in the amount of AOX in effluent waters from 5-7 to 1-2 kg/metric ton of pulp. Today the use of chlorine in the pulp and paper industry is continuously decreasing, in spite of an increase in the production of bleached pulp.

Oil Pollution

Petroleum hydrocarbons are released to the Baltic Sea from many sources. These include river and land runoff, direct discharges from cities and industries, and atmospheric deposition. The cleaning of oil tanks and other deliberate releases from ships, as well as oil spills resulting from accidents are examples of other important sources. A new threat, which may increase in the Baltic region in the future, is the growing impact of releases connected with oil prospecting and off-shore oil platforms. Germany is already extracting oil from platforms in its region and further prospecting is being conducted by several countries in Kattegat and the Baltic Proper.

It is difficult to make even a rough estimate of the total load of petroleum hydrocarbons on the Baltic Sea, because the quantities released from the various sources are virtually unknown. According to an estimate made by the Helsinki Commission in 1986, however, it seems that the input from rivers and atmospheric deposition accounts for more than 50% of the total load, while the contribution made by ship accidents and deliberate oil discharges from ships is relatively low. The total amount of oil products released into the Baltic Sea was estimated to range from 21,000 to 66,000 metric tons per year. The petroleum content in the surface water is comparable in all parts of the Baltic Sea but, compared to the North Sea and North Atlantic, the levels are two and three times higher, respectively.

Environmental Policy and Cooperation in the Baltic Region [1, 3]

THE SPECIAL PROBLEMS OF THE BALTIC SEA

The Baltic Sea is considered to be one of the most environmentally threatened seas in the world. These threats consist of eutrophication, different kinds of pollution and toxins, and the extinction of different species. Pollution originates from several kinds of sources: land-based activities, vessels, sea-bed activities. Overexploitation of fish populations may also cause problems.

It is estimated that by far the most of the pollution in the Baltic Sea originates from land-based sources. However, regulation of these sources has proven to be more difficult than regulation of pollution from vessels. The most important convention on land-based pollution in the Baltic Sea is the Baltic Convention.

INTERNATIONAL CONVENTIONS FOR THE BALTIC SEA

A. Convention for the Prevention of Marine Pollution from Land-Based Sources ("Paris Convention")

> Signed 1974. In force in 1978. Sweden, Germany, Denmark and most states bordering the northeast Atlantic and the North Sea, except USSR. The Paris Convention applies to the North Sea and the northeast Atlantic Ocean. It also applies to the water between Sweden and Denmark. Deals only with land-based pollution sources. The Convention adopts both recommendations and binding decisions.

B. 1972 London Dumping Convention (LDC)

> Signed 1972. In force in 1975. Global convention ratified by about 60 states, including all states bordering the Baltic Sea. In contrast to the Baltic Convention, which contains a total prohibition on dumping, the LDC does not prohibit all kinds of dumping. Substances listed on the "black list" (organohalogens, mercury, cadmium, oil and high-level radioactive wastes) must not be dumped at all, while compounds from the "grey list" (arsenic, lead, organosilicons, cyanides) may be dumped after a prior special permit by the national authority.

C. 1973/1978 MARPOL Protocol

> Signed in 1973 and 1978. In force in 1983. Global treaty ratified by about 50 states, including all states bordering the Baltic Sea. Deals with pollution from "normal operations" of vessels: oil, noxious liquids in bulk, harmful substances in packaged forms, sewage and garbage. Within the Baltic Sea, stricter rules are applied.

D. Nordic Environmental Protection Convention (NEPC)

Signed 1974. In force in 1976. Denmark, Sweden and Finland.

E. ECE Convention on Environmental Impact Assessment in a Transboundary Context (EIA-convention")

Signed 1991. 26 state members of the Economic Commission for Europe.

F. 1979 Convention on Long-Range Transboundary Air Pollution

Signed 1979. In force in 1983. 30 state members of ECE, including all states bordering the Baltic Sea).

G. 1973 Convention on Fishing and Conservation of the Living Resources in the Baltic Sea

1973/1974. All states bordering the Baltic Sea.

H. Conventions on nuclear activities

I. Conventions on the conservation of endangered species

THE BALTIC CONVENTION

In 1974 the Convention on the Protection of the Marine Environment of the Baltic Sea Area (called the Baltic or Helsinki Convention) was signed in Helsinki by all countries bordering the Baltic Sea. This convention was regional and it included several types of pollution. Technically, it concerned land-based pollution, pollution from ships, dumping, combating (clean-up), control (monitoring) and assessment of the environment. The Convention formally entered into force in May 1980. The steering agency of the Convention is the Baltic Marine Environment Protection Commission - Helsinki Commission (HELCOM). HELCOM is an inter-governmental organization. The governments choose their representatives in the Commission, mostly experts from environmental protection agencies who meet every year in Helsinki. Since 1980, there have been ministerial meetings every fourth year [7].

HELCOM elaborates recommendations on different environmental threats almost every year. All the HELCOM recommendations are taken by all the states unanimously, not just a majority. The recommendations are only advisory and are not legally binding.

In 1990 a revision of the convention was initiated and in April 1992 the new version was signed by the parties. It is somewhat stricter than the one from 1974, for instance, concerning the stipulation that the best-available technology shall be used and that the Contracting Parties shall apply the *precautionary principle*, i.e., to take preventive measures when there is a reason to assume that substances or energy introduced directly or indirectly into the marine environment may create hazards to human health, harm living resources and marine ecosystems, damage amenities or interfere with other legitimate uses of the sea, even when there is no conclusive evidence of a causal relationship between inputs and their alleged effects.

The participants decided that a Baltic Sea Joint Comprehensive Programme should be elaborated in order to restore the Baltic Sea. In order to be effective, the program sets priorities and make recommendations for financing of actions in the catchment area as a whole and thus concentrate efforts where they are most needed. The joint comprehensive program encompasses both preventive actions to promote sustainable use of the Baltic environment, and curative actions to rectify the legacy of environmental degradation.

Through the feasibility studies, the program has identified the action needed to control pollution at 132 "hot spots." 98 actions were identified in Russia, Estonia, Latvia, Lithuania, Belarus, Ukraine, Poland, and the Czech and Slovak Republics.

The financial institutions participating in the Baltic Sea Conference are the European Bank for Reconstruction and Development, the European Investment Bank, the Nordic Investment Bank and the World Bank. Implementation of the entire long-term program is estimated to cost at least 18 billion ECU over a twenty-year period. The 132 "hot spots" comprise actions to address point and non-point source pollution in the Baltic Sea catchment area and are estimated to cost about 10 billion ECU.

References

1. Andreasson-Gren, I.M., G. Michanel, and J. Ebbesson, 1993: Economy and law: Environmental protection in the Baltic Region. In: Lars Ryden (ed.), *The Baltic Sea Environment*. Uppsala, Sweden: The Baltic University Programme, Uppsala University. Session 7. Benny Killinger Ord & Vetande AB.

2. Backlund, P., B. Holmbom, and E. Lepäkoski, 1993: Industrial emissions and toxic pollutants. In: Lars Ryden (ed.), *The Baltic Sea Environment*. Uppsala, Sweden: The Baltic University Programme, Uppsala University. Session 5. Benny Killinger Ord & Vetande AB.

3. Bergström, G.W., 1993: Environmental policy and cooperation in the Baltic region. In: Lars Ryden (ed.), *The Baltic Sea Environment*. Uppsala, Sweden: The Baltic University Programme, Uppsala University. Session 8. Benny Killinger Ord & Vetande AB.

4. Forsberg, C., 1993: Eutrophication of the Baltic Sea. In: Lars Ryden (ed.), *The Baltic Sea Environment*. Uppsala, Sweden: The Baltic University Programme, Uppsala University. Session 3. Benny Killinger Ord & Vetande AB.

5. Haokanson, L., 1993: Physical geography of the Baltic. In: Lars Ryden (ed.), *The Baltic Sea Environment.* Uppsala, Sweden: The Baltic University Programme, Uppsala University. Session 1. Benny Killinger Ord & Vetande AB.

6. Kautsky, L., 1993: Life in the Baltic Sea. In: Lars Ryden (ed.), *The Baltic Sea Environment.* Uppsala, Sweden: The Baltic University Programme, Uppsala University. Session 7. Benny Killinger Ord & Vetande AB.

7. The Swedish Council for Planning and Coordination of Research, 1994: *A Future for the Baltic? Scientists Discuss the Future of the Baltic Sea.*

IDENTIFYING SOURCES OF RADIOACTIVE AND HEAVY METAL CONTAMINATION IN THE CASPIAN SEA: FUTURE RESEARCH OPPORTUNITIES

PÁVEL SZERBIN
Joliet-Curie National Research Institute for Radiobiology and Radiohygiene
Budapest, Hungary

Introduction

The interest to extend the knowledge in the field of the mechanisms regulating the behavior of radionuclides and heavy metals in different components of the environment has been raised in the second half of the 20th century. The use of nuclear energy for peaceful and military purposes, the radioactive fallout dispersed by nuclear weapon tests and, more recently, the contamination due to the Chernobyl accident have been the most important sources of radionuclides in terrestrial and aquatic ecosystems. As a consequence of industrial activity the environmental distribution of naturally occurring radioactivity and heavy metals was changed, and new areas of elevated levels of these contaminants appeared. The problems of environmental pollution by radionuclides and heavy metals appeared simultaneously. The first large-scale nuclear accident at Kyshtym (Russia) in 1957 caused heavy radioactive contamination of populated areas, soils and surface waters. For a rather long time this accident could not be studied exhaustively because of the Cold War situation; but now the data are accessible to scientists and they represent a source of invaluable information. The first detection of cases of heavy metal contamination in Japan appeared in 1956 (Minamata disease), resulting in extensive research concentrated on the food-chain transport and toxicity of heavy metals.

The monitoring of radioactive and heavy-metal pollutants in terrestrial and aquatic ecosystems is indisputably one of the most important tasks of science. In recent years several projects were initiated by the European Commission (EC), the International Atomic Energy Agency (IAEA), and UNESCO which were focused on the subject of radioactive and heavy-metal contamination of large European rivers, such as the Danube and Dnieper [1, 2, 3].

The importance of these two large rivers is obvious; the Danube is the second largest transboundary river and the Dnieper was seriously affected by the

M.H. Glantz and I.S. Zonn (eds.),
Scientific, Environmental, and Political Issues in the Circum-Caspian Region, 245-254.
© 1997 *Kluwer Academic Publishers. Printed in the Netherlands.*

Chernobyl accident. The post-Chernobyl radiation monitoring case studies can allow us to extract the unique methodological experience and to disseminate it to other fields of environmental monitoring of water bodies. As a result of the studies mentioned above, many aspects of water quality assessment, monitoring, data collection and exchange technology, and other effects on Black Sea contamination are now clearer, but there are many tasks that remain to be undertaken in future joint research projects. The experience of the Ukraine, Danube riparian states and countries involved in international projects should be studied and implemented for other aquatic ecosystems.

Unfortunately, we do not have a comprehensive national or international study on the problem of contamination of the Caspian Sea from its inflowing river catchment areas. There is a need for such a study, since in the years of the so-called planned economy of the former socialist countries and the USSR several cases of large-scale environmental pollution by radionuclides and heavy metals took place. However, information on these events was not accessible by the public. Information on the present levels of radioactive and heavy-metal contamination and assessment of the consequences of such contamination for the population in the Caspian region, the catchment areas of the main tributaries are needed. The largest European rivers, the Volga and the Ural, should be examined using the international expertise of the radiation and heavy-metal monitoring of national services and research institutes.

Potential Sources of Contamination

NUCLEAR FALLOUT

Following the first nuclear weapons test at Alamogordo, New Mexico (USA) in 1945, radioactive contamination was found on X-ray films produced by an Eastman Kodak Company factory at approximately 1000 miles from the point of detonation. The subsequent studies on radioactive fallout suggested that not only the direct irradiation could be responsible for doses to living organisms but that the contamination from radionuclides could also play an important role.

The fallout from the ongoing atmospheric nuclear weapons testing until 1963 was more or less uniformly distributed throughout the world as a consequence of the high altitude of the radioactive cloud (Table 1). Differences were observed between the Northern and Southern Hemispheres, as exchange between the air masses of the hemispheres is limited (Fig. 1).

Aquatic ecosystems were contaminated by the radioactive fallout, as a result of direct deposition on the surface or by waters of the catchment areas. The rivers transport radioactivity that had been deposited directly onto the water surface or by runoff from the contaminated catchment.

TABLE 1. Deposition density in the 40°-50°N latitude belt form atmospheric nuclear testing [5]

Radionuclide	Deposition density* (Bq m^{-2})
Sr-89	20,000
Sr-90	3,230
I-131	19,000
Cs-137	5,200
Ba-140	23,000
Pu-238	1.5
Pu-239	35
Pu-240	23
Pu-241	730
Am-241	25

* total deposition until 1990 not corrected for decay

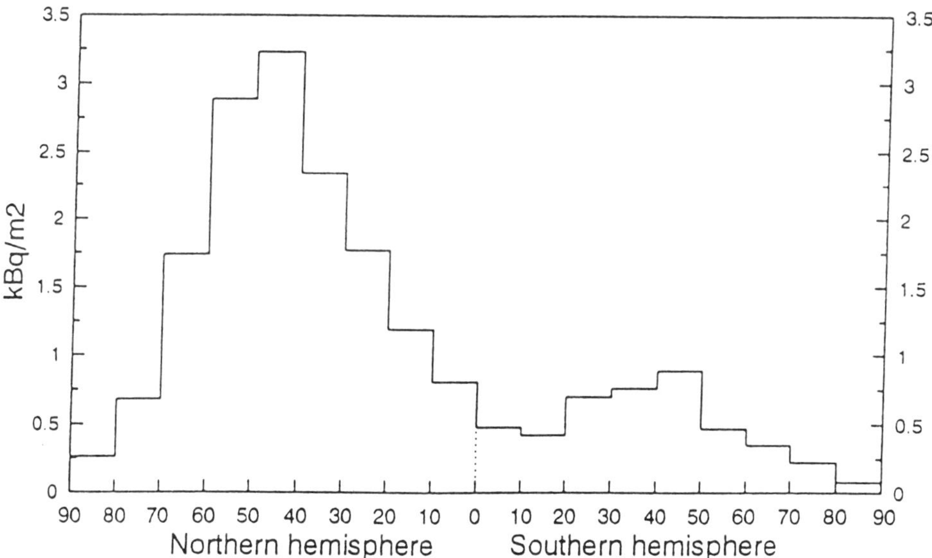

Fig. 1. Integrated deposition density of Strontium-90. Unit: kBq m^2 [5]. Cumulative deposition is 0.45 x integrated in 1995.

Accidents

Several accidents at nuclear facilities have contributed to environmental radioactive contamination on the territory of the former Soviet Union. Particular interest should be given to accidents with extensive temporal and spatial consequences. In 1957 a container of liquid nuclear waste blew up at a nuclear fuel reprocessing plant, Mayak-40, in the Chelyabinsk region in the southern Urals. The radioactive cloud was directed to the northeast and the catchment areas of the Techa and Iset rivers were seriously contaminated. The catchments of the Volga and Ural rivers were not affected by this accident.

The Chernobyl accident released a considerable amount of radioactivity to the atmosphere (Table 2). As a consequence of the extensive radioactive contamination of the drainage basins of the Dnieper and Pripyat rivers, a wide area was formed in the northern Ukraine, from which contaminated runoff could flow downstream through the Dnieper cascade system to the Black Sea. About 9 million people living in the middle to lower Dnieper basin received doses of radioactivity through direct consumption of water form the Dnieper River system and up to 30 million received doses through the contamination of irrigation water and fisheries.

TABLE 2. The estimated releases and deposition of long-lived radionuclides from the Chernobyl accident. All activities are decay-corrected to 26 April 1986 [7].

Radionuclides	Release in PBq*	Activity in PBq* deposited in European part of the USSR
Cs-137	~ 100	~ 30
Cs-134	~ 50	~ 15
Sr-90	~ 8	~ 7
Ru-106	~35	~ 25
Ce-144	~ 90	~ 75
Ag-110m	~ 1.5	~ 0.5
Sb-125	~ 3	~ 2
Pu-239, 240	~ 0.055	~ 0.05
Pu-238	~ 0.025	~ 0.02
Pu-241	~ 5	~ 4
Am-241	~ 0.006	~ 0.005
Cm-242	~ 0.6	~ 0.55
Cm-243, 244	~ 0.006	~ 0.005

*petabecquerel

Trench traps for the interception of contaminated suspended sediments were dug in the bottom of the Pripyat River, downstream of the Chernobyl Nuclear Power Plant and a number of filtering dams were constructed on rivers and streams. Radioactive wastes from the accident were buried in shallow sandy trenches without any attempt to protect groundwater from contamination. These, and most of other aquatic countermeasures, proved to be extremely expensive and inefficient in terms of dose reduction. The main cause of these failures was a poor knowledge of the key mechanisms governing radionuclide transport in aquatic systems.

The radioactive fallout from the accident caused considerable contamination of some areas of the Danube catchment. The catchment areas of Caspian inflow rivers were contaminated to a lesser extent, but the total area is very large. It is very important, since, for a long time after the deposition, the main source of contamination is the runoff from the soil surface of the catchment. Obviously, studies after the Chernobyl accident were concentrated on the Dnieper-Danube-Black Sea system. But almost nothing is known about the consequences of the Chernobyl accident in the Caspian area, and there are no available literature data on the behavior of radioactive contaminants in this important aquatic system. However, the considerable ecological, social, and economical importance of the Volga-Ural-Caspian Sea system should turn the interest of politicians and scientists to the problems of this ecological area.

INDUSTRIAL ACTIVITY, HOSPITALS, RESEARCH AND MILITARY BASES

Radioactive contamination could be of either nuclear or non-nuclear industry origin. The nuclear industry includes uranium mining and milling, nuclear power generation, reprocessing and radioactive waste treatment and disposal. Under normal operations of nuclear power plants or experimental reactors, the discharge of radioactive substances to the environment are regulated by the appropriate authorities. The releases are kept low (Table 3) in order to protect humans and the environment, and environmental contamination as well as the doses received from normal operation of nuclear power plants are, thus, in general negligible. The environmental impact of uranium mining and milling is more important, but the highest potential risk is from reprocessing and nuclear waste treatment and disposal. In the literature there are lots of data available dealing with this subject.

TABLE 3. Normalized average release from LWRs with liquid effluent in GBq (GWea)

	^{51}Cr	^{54}Mn	^{55}Fe	^{59}Fe	^{58}Co	^{60}Co	^{65}Zn	^{110m}Ag	^{134}Cs	^{137}Cs
PWR	2.8	1.9	5.6	0.22	27	7.6	0.016	2.4	2.3	4.6
BWR	21	5	1.9	1.7	2.8	11	1.3	0.06	1.3	2.9

LWR = liquid water reactor; PWR = pressurized water reactor; BWR = boiled water reactor; GBq = gigabecquerel; GWea = gigawatt energy production per year

The most serious cases of contamination from the nuclear fuel cycle in the former known Soviet Union are studied thoroughly. Thus, during the period 1949-1952 liquid and low-level radioactive waste was disposed directly into the Techa River system from the Mayak plant. About 99% of the released radionuclides were deposited in the river's flood lands and bed. Most of the radioactivity discharged to the Ob River system can, thus, still be found in the Techa River and in particular the reservoirs. However, some of the radioactivity must be elsewhere, probably in the world's oceans via the Kara Sea and the Arctic Ocean.

In connection with Mayak a number of reservoirs for the disposal of high-level radioactive waste has been established. One of them is the Lake Karachay, and in 1967 dust from the shoreline of the lake containing high concentrations of radionuclides was dispersed by the wind to a distance of 75 km. The huge amount of radioactive material in Lake Karachay makes this open-air waste repository a potential source of environmental contamination in the Urals.

There are no known cases of serious large-scale environmental radioactive contamination in the Caspian inflow river catchment area. However, many industrial enterprises involved in the nuclear production cycle are located in the Caspian Sea catchment area (e.g., Obninsk, Kalinin, Moscow, Arzamas, Melekess, Balakovo, Orenburg, Astrakhan —all of them within the Russian Federation). These sources have to be thoroughly studied from the point of view of environmental impact assessment identifying the potential risk to human health.

Military programs (e.g., the production of nuclear warheads) contribute to environmental pollution. Particular sources of contamination could result from the use of radioactive material in hospitals and in research and educational institutions. The impacts of these disparate sources are considered to be negligible compared to industrial sources. The non-nuclear industry releases to the environment are mainly natural radionuclides of uranium and thorium series (from the phosphate fertilizer industry, coal-fired power plants, etc.). As a result, in some parts of the ecosystems, elevated levels of radioactivity from naturally occurring radionuclides can be observed.

Particular attention should be paid to military bases. In the Central and Eastern European countries of the former Warsaw Treaty Organization, local authorities were faced with acute environmental problems on the sites of the evacuated military bases after the withdrawal of Soviet troops. The soils and the groundwater were seriously contaminated by heavy metals and other chemicals, mainly products of the oil industry. Unofficial information suggests that similar situations now exist at the evacuated nuclear missile bases in Belarus and the Ukraine.

Heavy metals appear in the environment mainly from industrial and agricultural sources. The Caspian Sea catchment area can be characterized by high concentration of ecologically dangerous industries such as ferrous and non-ferrous metallurgy, mining, chemical, cellulose and electric power production. The situation is aggravated by the widespread use of power-consuming and resource-consuming technologies, by a shortage of effective purifying installations, and by low effectiveness of recycling. As a consequence of unsatisfactory of water purification, most of the heavy metal contaminants are released directly to wastewaters. The wastewater-draining rivers transport the pollutants in soluble and suspended-matter forms, as a result, there are considerable amounts of heavy metals in the bottom sediments. The main pathway of the heavy metals to humans is the food chain, that is, via aquatic organisms and terrestrial agricultural production.

The main sources of mercury are as follows: the chemical industry, medical industry, cellulose industry and as a result of fossil energy production. The most important sources of cadmium are metallurgy, the chemical industry, battery production and waste combustion. Large amounts of cadmium are released to the environment by waste disposal and fertilizing. Lead, transported by rivers to seas and oceans, is about 75% of industrial and transport origin, while agriculture is responsible for 15% of the lead in rivers. Copper released to the atmosphere is mostly (i.e., 75%) the result of human activity. Most of the other heavy metal discharges to the atmosphere and to terrestrial and aquatic systems could be accounted for from industrial and agricultural sources.

Very few data are available in the open literature on the subject of heavy metal contamination of the Caspian region and the Caspian catchment area. Important future tasks of collaborative studies in this region could be as follows: an inventory of potential sources of contamination, an investigation of the current levels of the various heavy metals in the water ecosystems, a risk assessment to the public in the area and the identification and evaluation of countermeasures for the reduction of unacceptable risks resulting from elevated levels of contamination.

Results of Previous Radioactivity and Heavy Metal Studies

As previously mentioned, several studies were initiated to investigate the sources of Black Sea contamination from the Danube and Dnieper catchments. It was shown that impact of routine releases of nuclear power plants to the environment was negligibly small. Most of the anthropogenic radioactivity which was detected in the investigated areas originated from the Chernobyl accident. In the case of the Danube it was revealed that the level of contamination decreases gradually from the upper parts to the Black Sea. The waste-water releases of large cities in general did not influence the radioactivity levels of the Danube River and its sediments. However, at some sites "hot spots" of man-made radioactivity of unknown origin were detected (Fig. 2).

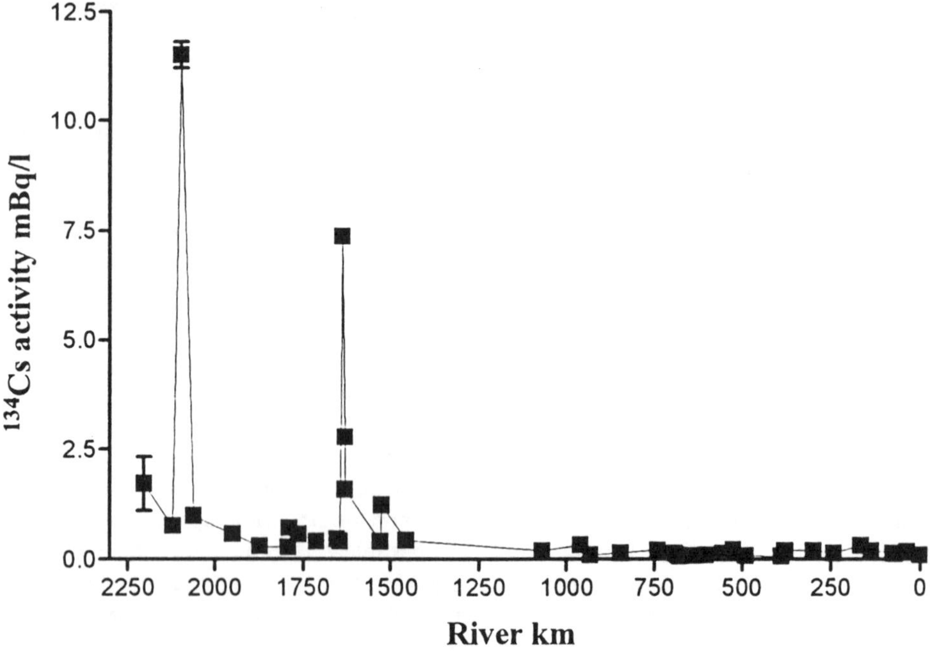

Fig. 2. ^{134}Cs concentration in Danube water from the German-Austrian border to the delta.

Regarding the main sources of elevated natural radioactivity in the region, analyses pointed to the mining industry, coal-burning power plants and metallurgical plants. The Danube tributaries, flowing through regions of high-density conventional industry, showed enhanced levels of naturally occurring radionuclides. It was concluded that the main source of radioactivity of the rivers is the runoff from the contaminated catchment. Long-lived ^{137}Cs deposited on the land surface after nuclear fallout from the Chernobyl accident is still in the upper layer of the soil. Erosion processes transport this radioactivity by runoff to the water bodies.

One of the main lessons of the Danube study was that adequate international monitoring of transboundary rivers and other large aquatic or terrestrial ecological regions was lacking and that there was a strong need for standardized and thoroughly intercalibrated methods. The main activity of the monitoring networks should be to obtain regular reliable and comparable data on environmental contamination.

Investigation of the heavy metals in the Danube showed that concentration of the contamination frequently exceeded the standards for safe drinking water. During the last decade, heavy-metal pollution of the Danube decreased. It was shown that the highest contamination levels were found in the smallest particle-size fraction of sediments. Most of the metals in sediments exceeded the lowest natural values, indicating that there was anthropogenic input above the natural background.

Mainly, the tributaries were polluted, but along the Danube sediment was also influenced by pollution form downstream tributaries.

The behavior of different classes of heavy metals and other toxic elements and substances, in many cases, is comparable with the main mechanisms of physical and chemical changes in the catchment runoff, riverain and marine environments, with regard to suspended particles and with bottom sediments and pore water. Therefore, all methodological findings from radionuclide behavior studies can be applied to key parameter studies for heavy-metal transport by water pathways. This should lead to the development of an integrated radionuclide and heavy-metal monitoring system and could yield adequate answers about the dispersion of these pollutants in water ecosystems. It will also help us to assess the human health risks from these sources.

Research Opportunities

Most of the international research projects in the field of radioecology have been directed to the assessment of the consequences of the Chernobyl accident. The European Commission's (EC) Experimental Collaborative Project No. 3 [2] was concentrated on the investigation of the transfer of radioactive material from terrestrial ecosystems to and within water bodies, but was restricted to the area around Chernobyl. The Third Framework Project of the EC was opened for joint research among central and eastern European countries by the PECO project, but the financial support was not enough for a large-scale study. The EC Framework Program IV has been announced but as yet has no possibility for eastern European countries to join it. It was previously mentioned that PHARE (Poland and Hungary Aid for Reconstruction of Economy) has subsidized projects for Danube research, but the activity of the PHARE is restricted to the central European area. The EC's INCO-COPERNICUS program for cooperation with Third World countries and international organizations contains a particular and promising research direction: "endangered ecosystems." According to its information package, proposals for financial support directed to the study of environmental protection in the Caspian area are requested. By the deadline of the first call for proposals in the field of radioecology research about 400 proposals were submitted. However, the total budget was enough to support only 10 proposals. Presumably, the same situation prevails for other environmental pollution research projects. The newly independent states are still not able to support comprehensive studies in ecology and the budgets of international joint projects are quite limited. Consequently, the initiation of adequate environmental scientific research for the Caspian region may have little chance for success at the moment.

References

1. Aarkorg, A., 1995: Inventory of nuclear release in the world. In: NATO Advanced Study Institute Report, Advanced Course on Radioecology, Zarechny, Russia.

2. European Commission (EC), 1996: Modelling and study of the mechanisms of the transfer of radioactive material from terrestrial ecosystems to and in water bodies around Chernobyl. Experimental Collaboration Project No. 3. Final Report. EUR 16529. London, UK: European Commission.

3. PHARE (Poland and Hungary Aid for Reconstruction of the Economy), 1995: Impact of radionuclides in surface waters and sediments in the lower part of the Danube basin. Final Report, Phase 1. PHARE EU/AR/103/91. Brussels, Belgium: PHARE.

4. Lázló, F., and I. Liska, 1994: Organic and inorganic micropollutants in the Danube River along the Slovak-Hungarian border. In: *Environmental Monitoring Problems of Transboundary Rivers*, NATO ASI Workshop. Berlin: Springer-Verlag.

5. UNSCEAR (United Nations Scientific Committee on the Effects of Atomic Radiation), 1982: *Sources and Effects of Ionizing Radiation*. New York: United Nations.

6. UNSCEAR, 1993: *Sources and Effects of Ionizing Radiation*. New York: United Nations.

7. WHO (World Health Organization), 1989: Health hazards from radiocaesium following the Chernobyl nuclear accident. *Journal of Environmental Radioactivity*, **10**, 257-295.

VI. Regional and International Organizations

THE RISING WATERS OF THE CASPIAN SEA AND POSSIBILITIES FOR INTERNATIONAL COOPERATION

ILTER TURAN
Department of International Relations
Koç University
Istanbul, Turkey

The Problem

If the waters of Lake Leman were rising 10 cm a year, the riparians, France and Switzerland, would not find it very difficult to cooperate in searching for ways to deal with the problem. Why? Several reasons may easily be identified. First, there has been a long established and very stable arrangement regarding almost all questions of an international nature pertaining to the lake between France and Switzerland. Furthermore, there is no major set of outstanding questions between the two countries in areas other than the lake. France and Switzerland have been in existence as independent states, as full members of the international community of nations for a long time, whatever major conflict-laden issues may have existed between them have been resolved a long time ago, giving way to an overall stability in their relationship.

Second, both parties feel that they have an egalitarian relationship in more ways than just the legal sense. Although there exist differences between the power and capabilities of the two societies, each has access to sufficient national and international resources such that neither party would feel that the other would be able to impose its will on this or any other matter and gain an undue or an unacceptable advantage against the other.

Third, the relationship between the two riparians is characterized by high levels of mutual trust and confidence. Neither of the two countries feels that the other party has motives other than cooperating in addressing a problem which, by its nature, is a shared problem. The element of trust also implies that the limits or the range of what the other party may be expected to do and what the means are that it is expected to employ are predictable. It is known in advance, for example, that neither party will resort to the use of coercion or subversion in the relationship. It is also known that both parties will implement the commitments which they have undertaken.

M.H. Glantz and I.S. Zonn (eds.),
Scientific, Environmental, and Political Issues in the Circum-Caspian Region, 257-262.
© 1997 *Kluwer Academic Publishers. Printed in the Netherlands.*

Fourth, the two countries would be able to isolate the problem of the rising waters of Lake Leman from other problems that they may have between them, and treat it on its own merits, without fearing that it would have implications for other aspects of their relationship. That is, neither party would feel constrained that if they cooperated in dealing with the rising waters problem, inferences could be made from that behavior about other issues pertaining to the status of the lake such as the exploitation of its resources.

Finally, French cooperation with the Swiss may be facilitated considerably by the fact that whatever economic potential the lake offers is exploited to the mutual satisfaction of both riparians. The fact that there appear to be no major resources in Lake Leman that promise a considerable infusion of wealth to whichever country that exploits it, may render less significant the stakes involved for the party which tries to achieve a dominant position against the other in the lake.

I picked the imaginary problem of the rising waters of Lake Leman in order that it may serve the analytical purpose of offering a contrast to the conditions that we encounter around the Caspian which may render cooperation on the rising waters between the riparians somewhat difficult. Let us take up the contrast point by point. First, the relations between the riparians cannot be characterized as being stable and patterned both in general, and specifically with regard to the Caspian Sea. For example, three of the five riparians are newly independent states. The legal status of the sea is ill-defined, and even includes a legal debate as to whether the Caspian Sea is a sea or a lake. There are legal vacuums in some areas and significantly different interpretations of the law or legal precedents in others. Furthermore, the relations between the riparians is in a state of flux in almost all aspects. How the newly independent countries will relate to each other, what kind of broad international arrangements will eventually emerge to regulate the relationships between the Russian Federation and the successor states to the former Soviet Union in the economic, military and political fields are still evolving. There is a general feeling that the current situation is transient, and what lies ahead is not easy to predict.

Second, the relationship between the riparians of the Caspian Sea cannot be described as egalitarian. The riparians, three of which have only recently become independent, continue to be heavily reliant on cooperation with the Russian Federation to procure some of their basic needs and to have access to international markets. The integration of their economies to that of the Russian Federation deprives them of the means which would make them feel that they are party to an egalitarian relationship. Their military capabilities are insignificant in comparison to that of the Russian Federation. They have multi-ethnic populations that include Russians. They feel less than confident that these populations will maintain their commitment to the existing political arrangements in the long run.

Third, the relationship between the countries around the Caspian can hardly be characterized as possessing high levels of trust and mutual confidence. The two historical actors on the sea, Iran and Russia (later the Soviet Union) have had a checkered history of relations in which the lack of trust in each other's motives has been understandably typical. The relations between the newcomers and the historical actors have also been highly problematical. There are indications that some riparians have been involved in bringing about change of personalities in the politics of their neighbors. Some suspect that there are attempts of one country to export its political ideology to the neighboring states. Some riparians have accepted and implemented policies under pressure and duress which they would otherwise not have done.

Fourth, since there are many issue areas between the riparians where agreement is lacking, there is apprehension on the part of some of the riparians that what is done in one area may have strong implications for another area. Taking as an example riparian cooperation to deal with the problems generated by the rising waters of the Caspian while ignoring other issue areas where substantial disagreements seem to exist, may be viewed by the newly independent states as yet another step confirming the status quo which, not surprisingly, favors the position of the historical actors in the Caspian Sea basin.

Finally, the rich resource base of the Caspian Sea has rendered all actors very sensitive to how the various issues regarding internal waters, territorial waters, contiguous zone, exclusive economic zone and delineation of the continental shelf are settled. The economic potential of the sea, the plans of many riparians to exploit these resources so as to insure their own rapid economic development, the interest the world has shown in the development of these resources and their reaching the world markets, are some of the factors which render dealing with Caspian problems, including that of rising waters, difficult. For example, the rising waters mean that the sea surface area is getting larger and that it is moving inland. This means that the ten-mile line which is sometimes used to delineate the exclusive economic zone is receding as a result. Therefore, one cannot address problems of the rising waters without taking into consideration the economic implications of the changes this is causing in the utilization of resources of the sea. There are several interrelated problems the existence of which have to be recognized and addressed in dealing with any single problem pertaining to the Caspian. These include the legal status of the body of water (as a sea or as a lake), the questions of access of the riparians to the high seas, pollution accountability, fishing rights, navigation, sea borders, among others.

Conditions of Cooperation

If international cooperation is desired in order to arrive at viable solutions to a problem or a set of problems in an area, then a number of conditions need to be met in advance of the commencement of cooperative efforts. To begin with, actors which have an interest in the area must feel that the existing situation is highly unsatisfactory and that some major problems cannot be dealt with under the existing arrangement, and that these are likely to get worse in the future if ignored now. It is usually the case that at a given point the status quo is perceived to be more advantageous by some of the actors and disadvantageous by others. It may further be the case that an indefinite state of affairs is more to the liking of some who feel that time might work in their favor or that they have the means to affect gradually an outcome which is more to their advantage. The critical factor promoting cooperation is that each actor must still feel that the current situation is costly, and that such costs exceed any benefits which may accrue to it now or in the future, if the existing situation were allowed to continue.

A second condition must be that those engaging in a cooperative effort feel that they each will be able to affect the outcome in such a way that no solutions which they deem unacceptable will be imposed on them against their will. This condition can usually be met in advance by specifying who will take part in the negotiations, the instruments through which negotiations will be conducted, how decisions will be reached, and how they will be implemented and other procedural questions.

A third condition is that the actors involved in the cooperative process must have some confidence that the results achieved or the solutions which are arrived at are sustainable. This confidence, in turn, may be a function of several factors. For example, actors may feel more confident that an arrangement will be sustained, if they judge that all parties have gained sufficient benefits such that each one has a stake in observing or implementing it. Or, actors may feel that an arrangement is likely to stick if parties other than the riparians also have an interest, or have a commitment in making it work. Alternatively, confidence may develop, if actors demonstrate to each other that they will keep to their commitment by doing so in domains of their relationship which are not related to the issues at hand. Gestures of goodwill toward each other will often facilitate building mutual confidence.

Looking at the Caspian problems in light of the conditions of successful cooperation, it is not immediately apparent that all actors find the existing situation intolerable or highly costly. Historical actors feel that the existing arrangement works to their advantage and that others are not in a position to pressure for solutions which would redefine their place in the Caspian. Yet, a careful examination of the potential benefits of cooperation must be conducted, because future losses deriving from an indefinite state of affairs may be substantial.

The second condition may be met by specifying the organizational structure and the rules by which negotiations are to be conducted. Some proposals have already been put forward by the newly independent riparians of the Caspian Sea which could be taken as a beginning point.

The third condition appears to be the most difficult one to be met at this state, because the situation appears to be in flux not only internationally but also as regards the domestic politics of the riparians. It appears that an indefinite state of affairs exists also in the domain of foreign relations. To cite an example, the past electoral achievements of Mr. Zhirinovski, the position of the former Communist Party to restore the USSR and a relatively recent decision of the Duma to the same effect, may create understandable apprehensions among the other riparians such that their survival might be threatened. Some formulas which insure greater international commitment under the framework of the Organization for Security and Co-operation in Europe (OSCE) may be contrived, and other formulas may also be entertained, before attempting to implement cooperative solutions.

Strategy of Solution

In international relations, when two or more countries are faced with a set of interrelated problems to which solutions are difficult to devise, several different strategies may be used. Each strategy has its own merits. We may briefly review each of them:

Strategy 1: Start with less problematical issues and, as you solve them, move on to more difficult ones. The logic is that by starting with less problematical issues, the parties will reduce the number of problem areas, develop methods of working together to address mutual problems and develop mutual confidence so that they can tackle more difficult problems.

Strategy 2: Begin with the most difficult problem, work hard on achieving a solution. The dynamism generated by the solution of a major problem will facilitate the solutions of problems of a lesser order.

Strategy 3: Use a package approach. Especially if the set of problems are interrelated, it may be impossible to isolate the effects of what happens in one problem area from the other. In this case, adopting a package approach may be useful since the parties would be able to make concessions in one area in return for something that they may be accorded in another.

Strategy 4: Focus on one problem which the parties find important and feel that a solution may be possible. This does not mean that the existence of other problems will not be acknowledged. It simply means that the parties recognize an area where they can work together while other problems wait to be resolved.

In the case of the Caspian Sea, there is probably a hierarchy of importance of issue areas. The determination of the status of the body of water and the delineation of sea borders heads the list. Therefore, it may be possible to try to tackle the big problem first (Strategy 2). If the big problem is solved, then some other situations may cease to be problems, others can be solved by the momentum generated by the solution of the big problem. Since the problem areas are related, a package approach (Strategy 3) is also possible, but because the order of magnitude of the problems is vastly different, the potential for give-and-take across problem areas seems to be limited. Strategy 4, isolating the problem of the rising waters of the Caspian from other problems, is a possibility which requires further investigation. I judge, however, that the potential for success in one area, without addressing the major question, may be limited.

STRATEGIC PLANNING FOR THE SCIENTIFIC AND TECHNICAL COOPERATION AMONG THE LITTORAL STATES OF THE CASPIAN SEA

ABBASGHOLI JAHANI
Ministry Of Energy
Tehran, Islamic Republic Of Iran

Introduction

With the dissolution of the former Soviet Union and the creation of the newly independent states along the coast of the Caspian Sea, ever-increasing attention has been focused onto this important and unique body of water and its basin. A brief review of the potential and vast possibilities of the Caspian Sea and its basin in supplying the different needs and resources of its littoral states may give an indication of its vital importance for the economies of the aforementioned communities.

Based on preliminary assessments to date, about one third (1/3) of the total of economic products, one fifth (1/5) of total agricultural products and one third (1/3) of the hydro-electric energy of the former Soviet Union were supplied by the basin area of the Caspian Sea [1].

Considering the interconnections between the Caspian Sea and the Black Sea, the Azov Sea, the White Sea and the Baltic Sea, as well as its very considerable capacities in supplying livestock resources, and sturgeon, source of the world's finest and most famous caviar, make the role of the Caspian Sea and the rivers flowing into it very important. Freight transportation is another noteworthy feature of this important sea. Furthermore, this unique body of water along with its basin is also endowed with rich resources of gas and oil fields; different fields are either being presently exploited or being explored [1].

Preliminary assessments have proved that, if the necessary investments are made and sufficient exploration is carried out, vast gas and oil resources can be exploited in future decades [1].

Briefly, the Caspian Sea can be considered a very important area for natural resources, economics, social, environmental, political and strategic reasons.

M.H. Glantz and I.S. Zonn (eds.),
Scientific, Environmental, and Political Issues in the Circum-Caspian Region, 263-274.
© 1997 *Kluwer Academic Publishers. Printed in the Netherlands.*

With the breakup of the Soviet Union, the number of littoral states of the Caspian Sea increased from two to five. Hence, planning for the most suitable use and exploitation of undersea resources, and for protecting the environment of this unique body of water and securing conditions for sustainable development of this area, have become quite complicated. Planning requires the creation of suitable institutional and legal structures for the realization of effective inter-governmental cooperation among the coastal states.

Although political leaders, governmental and non-governmental organizations and even scientists and other experts from these littoral states, as well as organizations from around the world have, since the early 1990s, focused their attention on this important topic, major efforts have taken different forms. However, it appears that this region still lacks a strategic plan for a suitable organization for cooperation among the coastal states.

In this chapter two planning scenarios, i.e., downward and upward, are examined for developing a strategic framework and, after assessing their strong and weak points, the upward scenario has been recommended for reasons that follow.

Finally, recommendations have been made to create a suitable framework for effective cooperation among the littoral states, with an emphasis on the water balance of the Caspian Sea sea level fluctuations and the protection of its coastal zones.

Examination of Scenarios

Following the dissolution of the Soviet Union, two implicit schools of thought have governed the way scientific and technical cooperation among the coastal countries of the Caspian Sea have been developing. A brief explanation on each of them follows:

First Scenario

This view, that of initiating, developing and strengthening of regional cooperation, requires an official authority or organization at the highest possible level. According to this view, a central body is formed at the top of a pyramid. Then, all cooperation of different dimensions are placed at the base of this pyramid, thereby creating a structure based on a top-down (e.g., downward) movement.

In fact, the need for the existence of a coordinating authority to start any kind of activity forms the essence of this view. In this regard, mention can be made of the efforts that have already taken place to create the "Littoral States Cooperation Organization." In spite of initial agreements, whose framework is reflected in the Tehran Declaration of 1992, there is still no clear indication of the formal creation of this Organization, even five years after the Tehran declaration [2].

Following the delay in the formation of the Littoral States Cooperation Organization, international organizations and some of the littoral countries have started major activities, thereby, proposing an alternative to the realization of regional cooperation.

According to this alternative, the United Nations Environment Programme (UNEP), in lieu of this cooperative organization and on behalf of the littoral states, was made responsible for the coordination of the Caspian Sea environmental program [3].

Similarly, in a workshop which was set up in Paris from May 9-12, 1995 on the initiative of the International Atomic Energy Agency (IAEA) and with the cooperation of UNESCO and the Intergovernmental Oceanography Commission (IOC), the formation and preparation of a comprehensive multidisciplinary program was discussed. The preliminary framework for the proposed programs is reflected in the workshop report [4].

Evidently, the realization of each of these activities and the creation of the coordinating authorities placed at the top of the pyramid will be a very important and major step. However, the creation of this authority in such a way as to be able to organize the required activities efficiently, progressively and successfully will require the realization of the following two very essential preconditions:

a) Should the intended coordinating authority be a completely new structure (like the Caspian Sea Littoral States Cooperation Organization), it will have to fulfill all the legal and official procedures required for its creation. This will ensure the support and cooperation of the relevant governments for a timely execution of the Organizations's programs. It will also provide sufficient financial resources, funds and manpower.

   However, considering the complicated nature of existing problems related to the legal regime of the Caspian Sea and other regional economic, political and social issues, it does not appear to be possible to expect its realization in the short or even medium term, in spite of all the efforts that have taken place over the past five years.

   If, as intended, an international organization is to fulfill the responsibilities of the said authority, then, first there must be a complete understanding between all the relevant international organizations (of which there are quite a number) involved in problems related to the Caspian Sea, such that all said organizations will accept the selected organization as the coordinating body. Secondly, this international authority which is intended to replace the body that is to be organized, must be recognized by the governments of the littoral states. Due to the complicated nature of the various regional problems, it would seem unlikely to succeed. At the least it would take a very long time to resolve the issues involved and to create a suitable setting for such a coordinating organization.

b) Within the institutional structures of the governments of the littoral states, there are many government agencies and non-governmental organizations whose activities are related to the Caspian Sea.

Based on preliminary investigations, these internal bodies have different view points on similar matters. Hence, for the coordinating authority, described under (a) above, for it to be successful depends deeply on the level of coordination between governmental and non-governmental organizations within each of the littoral states. In the absence of such a national level coordinating body within these countries to coordinate national viewpoints and convey them to the inter-state coordination authority, the proposed international coordinating organization cannot be expected to succeed. The general structure of the coordination and management of activities in this scenario is shown in Fig. 1.

SECOND SCENARIO

In the upward scenario there is no need to create an intergovernmental coordination authority or even to create internal coordinating bodies within the littoral states of the Caspian Sea (see Fig. 1). There are already similar organizations within these states that, without having to go through two coordination levels, can directly communicate with each other and can prepare cooperative plans and programs. Some of the major advantages of this scenario are as follows:

a) Due to specialization, cooperation and joint projects get started more rapidly and progress faster.

b) The possibility of cooperation among specialized international organizations is more likely with this type of structure.

c) Since there is no major need for coordination among internal organizations, the funding requirements for the execution of projects can be met more easily.

d) The possibility for such cooperative bodies, if properly formed and organized, to join with national coordinating bodies and the inter-governmental coordination organization is perfectly feasible (see Fig. 1).

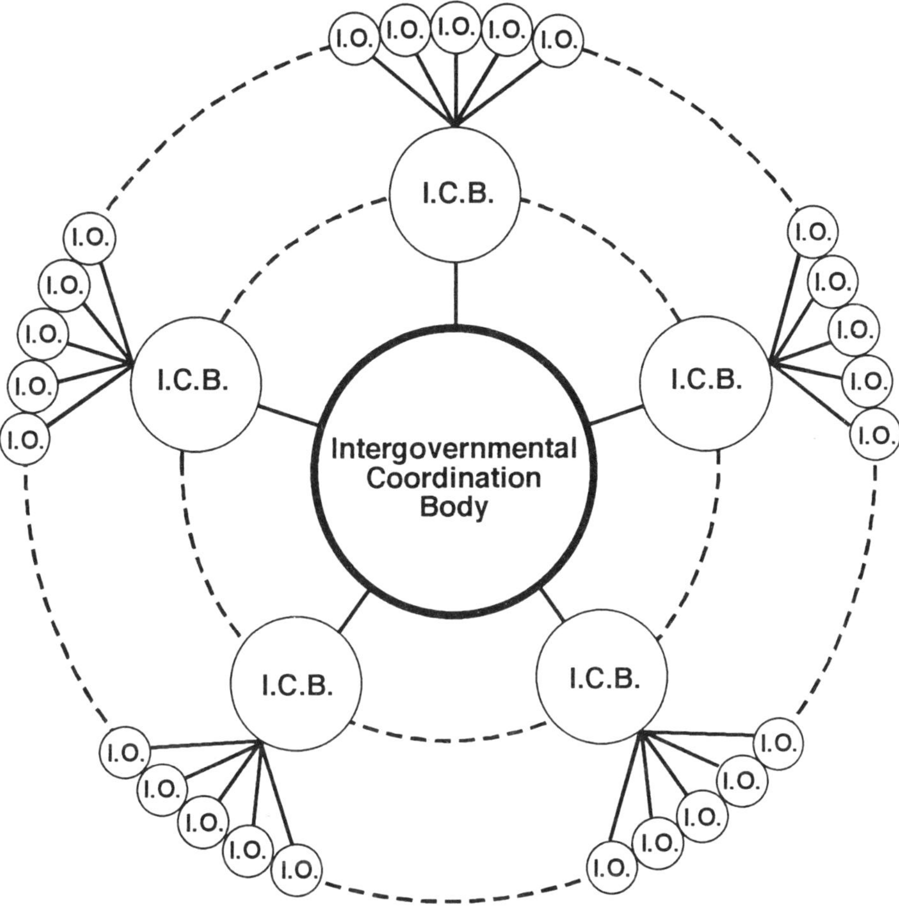

Fig. 1. Structure of first scenario. I.C.B. = Internal (National) Coordination Body; I.O. = Internal (National) Organization

e) In the upward scenario, reaching the completion level quickly in joint projects, due to the nature of cooperation involved, will encourage the littoral states to create an intergovernmental coordinating structure at the earliest possible time.

Figure 2 shows the general structure of this scenario.

The Second Scenario and Water Problems

With regard to the Caspian Sea, matters related to water can be considered from many different perspectives that exist among and within the coastal countries. Depending on their institutional and governmental structures, many organizations and ministries have varying responsibilities about aspects of Caspian water resources. For example, in the Islamic Republic of Iran, the responsibilities of the Ministry of Energy in matters related to the Caspian Sea, consist of the following two important matters:

a) Investigating the fluctuations of the Caspian Sea level, as well as predicting future water levels for more effective land-use planning of the coastal area.

b) Study, design and build suitable structures for the protection of the coastal zone.

In order to undertake these responsibilities the Ministry of Energy will have to organize the necessary steps, such as coordination with other ministries and organizations.

Furthermore, should any data and information on these topics be required by any of the internal and national agencies and institutions, it can be supplied only by the Ministry of Energy.

Based on the results of preliminary investigations, different institutional structures were tried in the newly independent nations, after the dissolution of the Soviet Union. However, after initial attempts, these structures tended toward stability. Presently, there are certain organizations within the littoral states whose responsibilities are similar to those of the Ministry of Energy. For instance, in Russia the Water Management Committee performs approximately the same duties as the Ministry of Energy in IR Iran, with regard to the above mentioned two important matters.

Therefore, each of the same organizations or agencies in the five littoral states, can take steps for the creation of cooperation authorities and for the preparation of necessary programs, without requiring the two coordination levels both at the national level and the main intergovernmental coordinating organization.

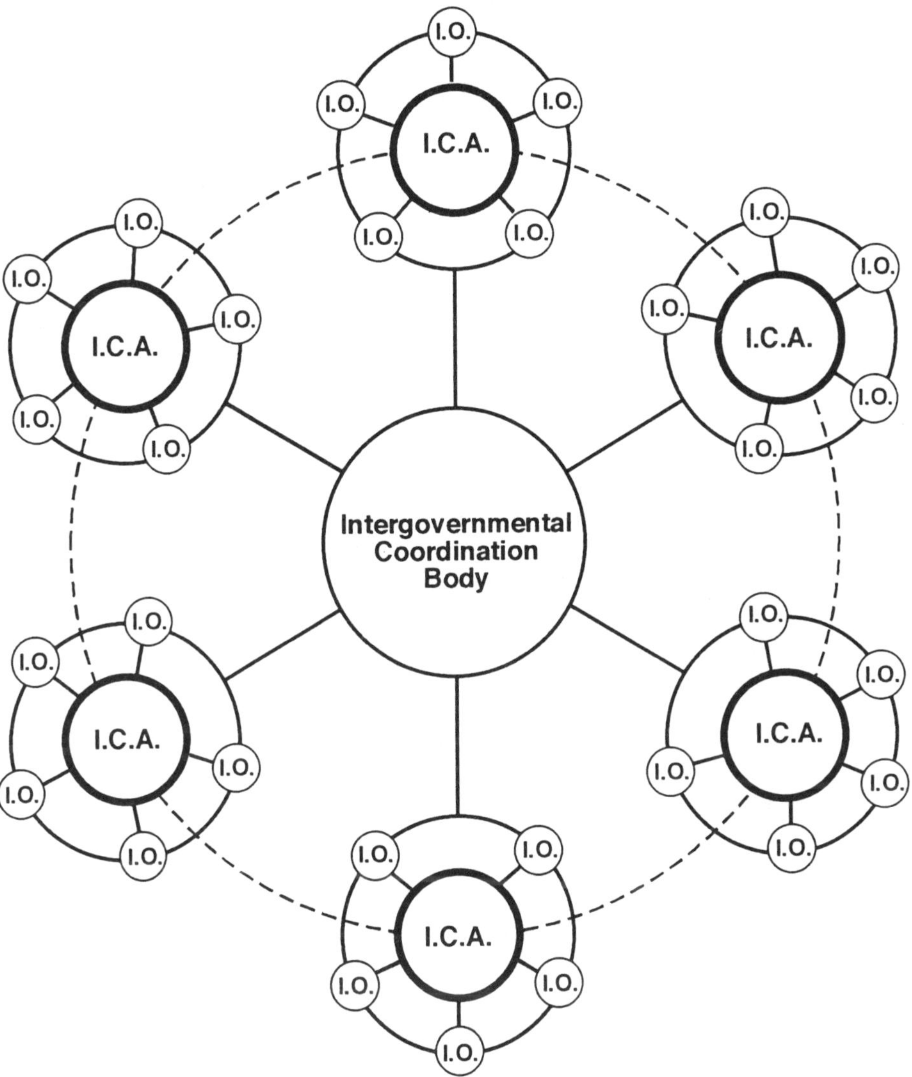

Fig. 2. Structure of second scenario. I.C.A. = Internal (National) Coordination Authority; I.O. = Internal (National) Organization

The creation of such an authority has been proposed by the littoral states in different forms. Russia, for instance, has proposed the formation of a "Committee for the Study of Water Level Fluctuations and Protection of Coastal Regions of the Caspian Sea." In this regard the draft constitution of the committee was prepared and reviewed in Makhachkala, Dagestan (Russia) in June 1994, where representatives of the littoral states participated. A "principal agreement" was also reached in this regard [5].

Similarly, the Islamic Republic of Iran has also suggested that an intergovernmental committee be created on matters related to water-level fluctuations and coastal protection of the Caspian Sea. The relevant draft constitution has also been distributed amongst the littoral states [6].

The main structure of this Committee is shown in Fig. 3. Not only can the formation of such a committee be the source for practical steps and extensive cooperation for the littoral states in the two areas mentioned earlier, but it can also fulfill one of the most important and urgent desires of these countries, namely, to produce a collective official document on the state of the Caspian Sea level fluctuations and to compile operational guidelines for defining suitable standards for land use planning in the coastal region. The activities and operations of this committee would be assisted and supported by a joint research center supported by joint investments from the littoral states, as well as by an appropriate data bank.

Based on agreements between the Russian Federation and the Islamic Republic of Iran, the creation of this research center with joint investments from both sides has already been approved by the parliament of the Islamic Republic of Iran. Similar steps are being taken presently by the Russian Federation. There are allowances and provisions made in these bilateral agreements and in the constitution of this center, for other littoral states to join [7]. Therefore, considering the temporary abolishment of the two coordination levels mentioned earlier, it seems that the creation of this committee is rapidly becoming a reality and cooperation among the littoral states is taking on a more practical and realistic form. It is certainly possible to take steps for the creation of other regional committees for other issues.

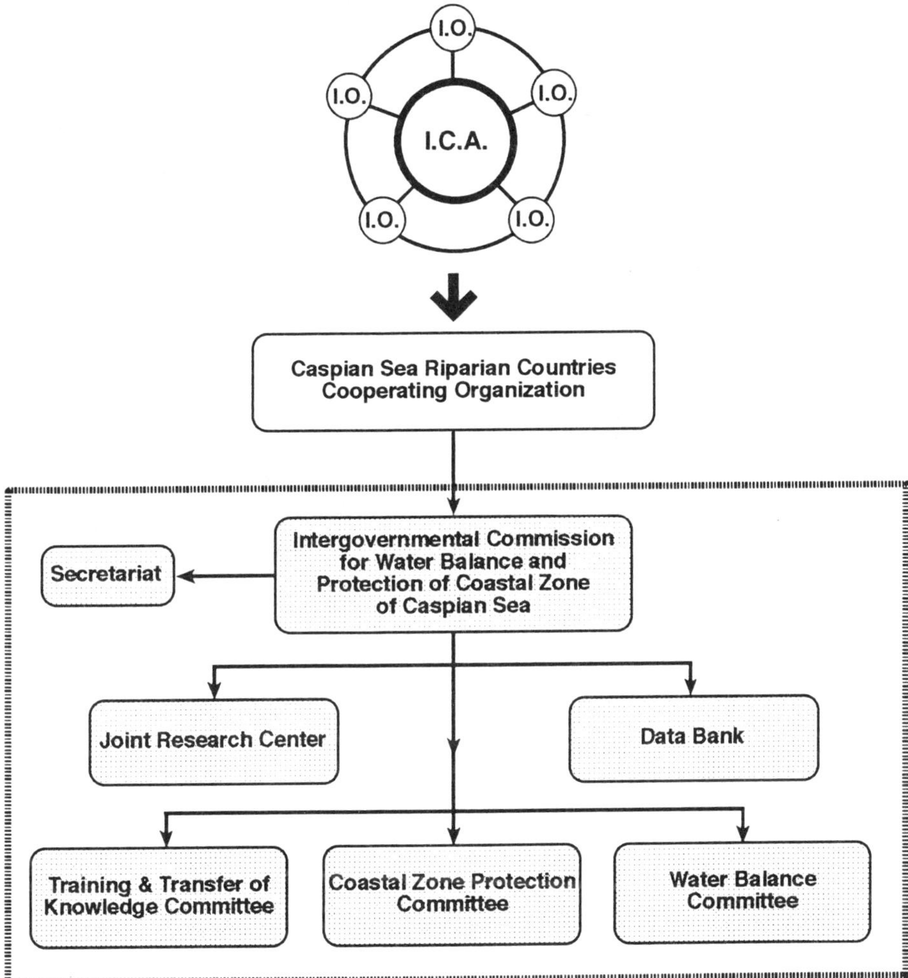

Fig. 3. Caspian Sea riparian countries cooperating organization

Fig. 4a. Scenes of destruction along the Iranian coastline resulting from Caspian sea level rise.

Fig. 4b.

Fig. 4c.

Fig. 4d. Scenes of destruction of Iranian coastline resulting from Caspian Sea level rise (cont'd.)

Conclusion and Recommendations

As an important body of water, the Caspian Sea plays a very crucial, sensitive and strategic role in the economic, political and social lives of the littoral states.

Organizing scientific and technical cooperation among the littoral states is considered as a very important step for the investigation of regional resources related to the Caspian. The basic and important precondition for the effective realization of such cooperation is the availability of a long-term strategic plan based on existing realities and potential trends. Two main scenarios for organizing cooperation were discussed in this chapter, i.e., the downward and upward structures. Although the downward scenario may be, due to the multidisciplinary and multisectoral nature of the cooperation, a very positive scenario in creating coordination and integration, it will, in practice, be impossible to create such a structure in the short run, because of different reasons such as political and economic incentives.

It is recommended, therefore, that steps be taken to activate the cooperative institutions among organizations responsible for certain specialized subjects by reducing the number of levels and the route of coordination. Provisions must be made to create such organs enabling them to join the main coordinating organization. In this respect, the recommended organization of the "Intergovernmental Committee for the Water Level Fluctuations and Protection of Coastal Zones of the Caspian Sea" is very likely to be created.

References

1. Sergei N. Rodionov, 1994: *Global and Regional Climate Interaction: The Caspian Sea Experience*. Dordrecht, The Netherlands: Kluwer Academic Publishers.

2. The Tehran Declaration 1992 - Documents of the Meeting of the Representative of the Littoral States of the Caspian Sea. Tehran, Iran.

3. Report of the First Meeting of the Regional Task Team on the Implications of Climate Change in the Caspian Sea Region. Moscow, Russia: 20-22 May 1994.

4. Report of UNESCO-IOC-IAEA Workshop on Sea Level Rise and the Multidisciplinary Studies of Environmental Processes in the Caspian Sea Region 1995. Paris, France 9-12 May 1995.

5. Rules of Committee on the Study of the Water Condition and Protection of the Coastal Zone of the Caspian Sea, Makhachkala (Dagestan, Russia) Meeting: August 1994.

6. Draft Constitution of the Inter-Governmental Committee for Sea Level Fluctuations and Coastal Protection of the Caspian Sea, 1994. Prepared by the Islamic Republic of Iran: Tehran.

7. Constitution of the Caspian Sea Joint Research and Studies Center: Documents of the Meeting on the Workshop on Caspian Sea Matters Between the Federation of Russia and the Islamic Republic of Iran: Moscow, 19-25 February 1993.

PERSPECTIVES FOR REGIONAL COOPERATION ON MAJOR ENVIRONMENTAL ISSUES IN THE CASPIAN SEA BASIN

GERHART SCHNEIDER
United Nations Environment Programme
Water Unit
Nairobi, Kenya

Introduction

The Caspian Sea is the largest inland water body on Earth. Having had only the Soviet Union and Iran as riparian states, it is now surrounded by five riparians states: The Russian Federation, Azerbaijan, Kazakstan, Turkmenistan, and the Islamic Republic of Iran, since the dissolution of the former Soviet Union in 1991. Its huge catchment includes upstream areas of rivers flowing into the Caspian from Armenia, Turkey, and Georgia.

The Caspian Sea is a terminal (or endorheic) lake, without a direct link to the world ocean. Thus, its water balance is governed by the river inflow from within its catchment (primarily from the Volga basin), direct precipitation onto the lake surface, and evapotranspiration determined by the climatic conditions. Being a relict marine basin (together with today's Mediterranean, Black and Azov Seas, it was part of the Tethys Sea in the Tertiary period), it shows remarkable marine features: Its sheer size (having five times the surface area of the world's second largest lake, Lake Superior), salinity of about 12 g/l, fauna (seals, etc.), currents, and near-shore cold upwelling areas. This "dual character" of the Caspian Sea has contributed to difficulties encountered by riparian states in their quest to agree on its unsettled legal status.

The vast oil deposits discovered in the last century in the Baku area have brought the Caspian Sea basin into the focus of interest of transnational corporations (e.g., international oil consortia), riparian governments, and numerous international and bilateral agencies. Former agreements between the Soviet Union and Iran (Persia) on navigation and fisheries are no longer deemed sufficient to address the various problems encountered in recent years. On the other hand, the rich oil deposits beneath the Caspian seabed, together with the appearance of three newly independent states (Azerbaijan, Kazakstan, Turkmenistan) make difficult any agreements among the riparian countries about the nature and legal status of the Caspian Sea.

M.H. Glantz and I.S. Zonn (eds.),
Scientific, Environmental, and Political Issues in the Circum-Caspian Region, 275-281.
© 1997 *Kluwer Academic Publishers. Printed in the Netherlands.*

Major Environmental Issues

Major environmental issues include water pollution, sturgeon fishing, biodiversity, sea level rise and desertification.

(I) Pollution comes from land-based activities (e.g., resulting in a riverine/shoreline inflow of nutrients and pollutants) as well as from "off-shore" activities (e.g., from oil installations, ships and tankers). The Volga, which drains the Russian heartland and accounts for about 80 major cities and numerous industries, clearly supplies the major pollution load. About 12 km^3 of wastewater each year (at different levels of treatment) is discharged into the Caspian Sea from the Volga basin. From available 1991 data, it has been calculated that 98 kilotonnes (kt) of nitrate, 126 kt of nitrogen, and 5 kt of phosphates entered the Caspian through the Volga in that year. 1992 data yielded estimates of annual heavy-metal loads at 114 t of cadmium, 1600 t of copper, 297 t lead, and 5100 t of zinc through the Volga.

However, there are other major sources of pollution: near-shore settlements, cities, and industries releasing effluents directly into the Caspian Sea, or into smaller Caspian basin rivers. The sewage discharge per year from coastal settlements is estimated at 8 km^3, but much less data on coastal settlements is available than it is for the Volga Basin. Also, oil pollution has a very long history in the Caspian Sea. Oil installations, ballast water from ships and tankers, and accidental oil spills and leakages are viewed as important sources of pollution, especially with hydrocarbons. The concentration of hydrocarbons in coastal waters is high (around 100-300 microgram/l), with peak concentrations off the coast of Azerbaijan registering 200-600 microgram/l.

In 1989 four oil-spill accidents were recorded, resulting in more than 300 tonnes of oil being spilled into the sea. One of these accidents alone resulted in oil being spread over more than 200 km^2, with apparently no one taking responsibility for cleaning up those spills.

(II) The sturgeon fishery: Approximately 90 percent of the world's harvest of sturgeon is harvested from the Caspian Sea. Official catch statistics show a sharp decline in recent years with a large (unknown) unreported catch being poached. The lack of enforcement of fisheries regulations seems to have had negative long-term effects, as seasonal and areal fishing restrictions are violated, damaging the long-term viability of the sturgeon stock. One can also be assumed that fairly sizeable quantities of juvenile sturgeon may have been caught, which is ultimately detrimental to the population dynamics of sturgeon. Apart from destructive fishing methods, the sharp reduction is the areal extent of the sturgeon's spawning grounds in rivers and in wetland areas, as a result of dam construction, has contributed to their decline. For example, after the construction of the hydropower plant at Volgograd, the spawning grounds for sturgeon was reduced from 3600 ha to 430 ha. Additionally, the change in the hydrologic regime of the Volga, through controlled water release from the dam, caused delay in and a shortening of the spawning periods. Because its natural reproduction has been seriously hampered, roughly half of the fish fry is, nowadays, produced in hatcheries.

Taking into account fish species other than sturgeon, a sharp decline in total fish catch has been observed. In the mid-1930s a total fish catch of 300,000 tonnes was reported and by the mid-1970s the catch had dropped to 100,000 tonnes annually.

(III) Other biodiversity issues (apart from sturgeon): The Caspian Sea reportedly hosts around 400 endemic species, adapted to the prevailing brackish conditions. These are in the group of the "Autochthonous Caspian Fauna," among which are goby and clupeid fishes, crustaceans, bivalve and gastropod mollusks. The effects of fishing pressure and pollution should not be overlooked.

(IV) Water level rise since the late 1970s has inundated nearshore areas, making the relocation of settlements and industries necessary. According to most scientists, this is not because of global climate change, but the result of long-term hydrologic oscillations. This view is supported by evaluation of long-term time series which show that, currently, the Caspian Sea level is approaching its level of the 1930s (around −26.5 m). As a consequence of the drop in sea level over a few decades (down to −29 m in 1977, for which remedial measures had already been planned!), settlements, industries and oil installations were moved closer toward the receding shoreline. Now that the water level has been rising since the late 1970s, these buildings and sites are being inundated and some had to be relocated at great cost.

(V) Desertification is prevalent in vast areas on the shores of the Caspian Sea, especially in Kalmykia, Dagestan, and Turkmenistan. This process has reached crisis proportions in some areas, largely caused by inappropriate land use, such as the overgrazing by large herds of sheep and attempts to convert natural pastures into croplands as in Kalmykia.

While all those "environmental issues" receive a lot of interest from various outside agencies that are ready to assist, these new possibilities being explored to channel support for regional cooperation in the Caspian Sea basin under a new international funding mechanism - The Global Environment Facility.

A Possible Role for the Global Environment Facility?

The Global Environment Facility (GEF) has four portfolios: One portfolio on Climate Change and one on Biodiversity (this is the interim funding mechanism for both the Framework Convention on Climate Change and for the Biodiversity Convention); International Waters, and the Ozone Layer. Except for the ozone layer issues, all other environmental issues related to the Caspian Sea can be linked to GEF portfolios, as discussed in the following paragraphs:

1. Riverine/shoreline pollutants and nutrients inflow issues: even if territorial boundaries were agreed upon in the Caspian Sea by riparian countries, pollutants and nutrients would still move freely. Considering the large pollution load entering the Caspian Sea through the Volga and other rivers, there is considerable scope for a regional pollution control program under the GEF-International Waters portfolio. Also, controlling pollution from various oil installations and industries already in place would require cross-border cooperation, including, for example, the possible setting of effluent standards and setting minimum technology requirements, for instance, for sewage treatment plants and processes. In addition, as has been proposed on various occasions, a regional "oil spill emergency unit" could be established to deal with oil spills from tanker and other oil-related accidents. Taking into account the longstanding experience in this field of a country like Iran, there seems to be scope for an exchange of experience and knowledge within the region.

2. With regard to the sturgeon fishery, sturgeon is a highly migratory species, with feeding grounds off Turkmenistan, spawning grounds in the Russian Volga River, and fishing grounds all over the Caspian Sea. Therefore, fisheries management cannot be pursued by one riparian alone, but would have to involve all five riparians, if the sturgeon stock is to be preserved in the long term. This, to a certain extent, is also applied to other commercial fish species in the Caspian Sea, especially for the anadromous and highly migratory ones.

3. There are other biodiversity issues: Because of the strong interest in the sturgeon fishery, interest in other species, many of them endemic, has been less pronounced. However, these species are also part of a unique natural heritage, which merits preservation by the riparian countries. Although of much less economic interest than sturgeon, they may also yield economic benefits to riparian countries, such as tourism. As with sturgeon, those species can only be preserved in the long-term through coordinated activities of all riparian countries.

4. Whether due to climate change or not, the Caspian Sea level rise issue is of major interest to all riparian countries, because of the large economic losses incurred through flooding of settlements, industries, oil installations, etc. This issue seems easier to tackle at national and local levels rather than through regional cooperation. Nevertheless, the exchange of experience with other regions of the world prone to flooding, like The Netherlands, could be helpful in designing the most appropriate and most cost-effective measures. Also, experiences with planning and risk-mapping of flood-prone areas near the shoreline could be undertaken. The sea level rise of the Caspian Sea makes certain adaption and mitigation measures necessary. In principle they are likely to be similar to those which could become necessary in other parts of the world concerned about

climate change-related sea level rise. In this regard, the Caspian Sea would be an ideal study area for identifying and assessing methods of adaption and mitigation of the impacts of sea-level rise which many scientists believe is to be expected.

5. Desertification of lands in the Caspian Sea basin, although of major importance to Dagestan, Kalmykia, and Turkmenistan, cannot be easily addressed through a regional coordination mechanism. Even at national and local levels, agricultural land use is extremely difficult to influence, especially with regard to achieving long-term sustainability in the face of short-term economic gains. Therefore, although locally a crucial issue, scope for regional or international involvement (apart form research efforts and information exchange) is limited. Nevertheless, this component qualifies for GEF-support.

Thus, in principle the major environmental issues in the Caspian Sea basin are eligible for GEF support. However, GEF support is subject to certain conditionalities, which are elaborated in following section.

Conditionalities of the GEF

There are two major sets of conditionalities for GEF funding: (1) The proven commitment by the governments involved to tackle a regional or global environmental problem and (2) The concept of "incremental costs".

THE COMMITMENT OF RIPARIAN GOVERNMENTS

In the concept of the GEF-International Waters Portfolio (GEF-IW), proposed activities should be "country-driven," with the implementing agencies of the GEF playing just a supportive role. This means that the riparian governments should not only turn to the GEF for securing outside funding for regional environmental problems, but should show strong national commitments to achieving a lasting success. Normally, riparian governments would be required to sign an agreement among them, identifying:

a. The major environmental problems perceived by all of them.
b. The methods/means to tackle those problems.
c. An appropriate regional "institutional framework" to achieve their goals.

Also, it is not enough for riparian governments to agree to undertake together research, monitoring, exchange of data, building of a regional database, etc. While research and monitoring activities can constitute a substantial part of a GEF-IW project, the governments have to show tangible actions in the field. This, of course, can cause problems, especially when governments have major difference of opinion, as is the case in the Caspian Sea basin. One example of the type of agreement which can be concluded by riparian countries is the "Tripartite Agreement on Lake Victoria." This was signed by the Permanent Secretaries from Kenya, Uganda, and Tanzania and did not require ratification by these three East African Parliaments.

In addition to the "commitment to be shown by riparian governments," another major conditionality of the GEF is the need for "incremental benefits," because the GEF would only fund "incremental costs." This has far-reaching implications, because it means that any GEF-funded project will require a substantial input of additional international, bilateral, and riparian country funding.

The term "incremental benefits and costs" means the following: For an "international waters" project to be funded by the GEF, it would not be enough to be beneficial only at a national scale, but should yield incremental regional, if not global, benefits. For instance, if a major pollution source such as a major near-shore city or industry, the GEF could not co-fund a sewage treatment plant, if the pollution source was of importance to only one country. However, if this hypothetical pollution source adversely affected other uses such as drinking water or fisheries in another riparian country, such an investment would be eligible for GEF-funding. Again, the incremental benefits and costs would have to be calculated, and only those incrementals would be covered by the GEF. In the given example, that means that the bulk of the cost for our proposed sewage treatment plant would have to be funded through other bilateral, international, or national sources of funds.

Unfortunately, the concept of "incremental benefits and costs" is still evolving and to date is unclear and is surrounded by confusion. The calculation of "incremental costs" (e.g., the GEF-contribution to a project) in my view is done arbitrarily, and cannot easily be explained to, or replicated by, non-involved persons. That necessary "country contribution," together with the time-limitation of GEF-funding (in most project documents I have reviewed, a period for GEF-support of up to five years has been anticipated) have political implications: Lasting achievements can be reached, only if structures and activities of long-term sustainability are developed (as well as the long-term national and regional funding mechanisms for them!). If riparian countries agree to the condition of "regional cooperation" to get GEF-funding without real political commitments, there is absolutely no chance of improving water quality, fisheries management, and other commonly used resources in a large international water body like the Caspian Sea.

KEY POINTS

- Although there are considerable differences in viewpoints among the Caspian Sea riparian countries with regard to its character, legal status and resource use, opportunities for regional cooperation on major environmental issues in the region exist.

- The Global Environment Facility could be a supporting mechanism for regional cooperation in the Caspian Sea basin. Its "International Waters" and "Biodiversity" portfolios are of special relevance in this context, if credible commitments by all five riparian countries can be demonstrated.

References

1. Governments of Kenya, Uganda, and Tanzania, 1994: *Agreement on the Preparation of a Tripartite Environmental Management Programme for Lake Victoria.* Signed by government representatives on 5 August 1994. Deposited with the Secretary-General of the United Nations; also on file with author.

2. Babaev, A.G., 1996: Ecological, social and economic problems of the Turkmen Caspian zone. In this volume. Dordrecht, The Netherlands: Kluwer Academic Publishers.

3. Chuikov, Y.U., 1996: Problems of ecological safety of the Asktrakhan region due to the rise of the Caspian sea. In this volume. Dordrecht, The Netherlands: Kluwer Academic Publishers.

4. European Environment Agency, 1995: *Europe's Environment: The Dobris Assessment.* Copenhagen, Luxembourg: European Environment Agency.

5. Global Environment Facility Secretariat, 1996: *GEF Operational Strategy.* Washington, D.C.: GEF Secretariat.

6. Golitsyn, G.S., 1996: Overview: History and causes of Caspian sea level changes. In this volume. Dordrecht, The Netherlands: Kluwer Academic Publishers.

7. Kosarev, A.N., and E.A. Yablonskaya, 1994: *The Caspian Sea.* The Hague, The Netherlands: SPB Academic Publishers.

8. Saiko, T.A., 1996: Environmental problems of the Caspian Sea region and the conflict of national priorities. In this volume. Dordrecht, The Netherlands: Kluwer Academic Publishers.

9. Anon., 1996: Pipe dreams in Central Asia. *The Economist Magazine*, 4 May, 37-38.

10. Vidaeus, L., and G. Schneider, 1992: Regional problem for environmental management of Lake Victoria. Discussion paper prepared by a World Bank/ UNEP Reconnaissance Mission. Washington, D.C.: World Bank.

11. Vinogradov, S., and P. Wouters, 1995: The Caspian Sea: Current legal problems. *Heidelberg Journal of International Law*, **55**, No. 2.

12. Vinogradov, S., and P. Wouters, 1996: The Caspian Sea: Quest for a new legal regime. *Leiden Journal of International Law*, **9**, No. 1.

13. Voropaev, G.V., 1996: The problem of the Caspian Sea level forecast and its control for the purpose of nature management optimization. In this volume. Dordrecht, The Netherlands: Kluwer Academic Publishers.

14. Zonn, I., 1995: Desertification in Russia: Problems and solutions: An example in the Republic of Kalmykia-Khalmg Tangch. *Environmental Monitoring and Assessment*, **37**, 347-363.

DO INTERNATIONAL ORGANIZATIONS HAVE A ROLE IN ADDRESSING CASPIAN ENVIRONMENTAL ISSUES?

MIKIYASU NAKAYAMA
Faculty of Agriculture
Utsunomiya University
Tochigi, Japan

International Water Bodies Require Integrated Management

National interests among countries are likely to diverge when it comes to international water bodies. Given the international context, however, inefficiency caused by interdependent water uses cannot be resolved through a single government's policies. Upstream countries tend to see little benefit from increasing or maintaining the flow and quality of water for downstream countries.

More than 200 of the world's river basins are shared by two or more countries. These basins account for about 60 percent of the earth's land area. Fragmented planning and development of transboundary water bodies has been the rule. Although more than 300 treaties have been signed by countries to deal with specific concerns about international water resources and more than 2,000 treaties have provisions related to water, coordinated management of international river basins is still rare, resulting in economic losses, environmental degradation, and international conflict [5].

International conferences such as the 1992 Dublin Conference on Water and the Environment and the 1992 U.N. Conference on Environment and Development (UNCED) in Rio de Janeiro have stressed the need for comprehensive management of water resources using the river basin as the focus of analysis. Cooperation and goodwill among states sharing a drainage basin are essential for the efficient development and utilization of international rivers and groundwater aquifers. In order to fulfill their own economic goals, it is important that such states formally collaborate to exchange data, share water, preserve the environment, and generate development programs that are of joint interest and benefit [6].

M.H. Glantz and I.S. Zonn (eds.),
Scientific, Environmental, and Political Issues in the Circum-Caspian Region, 283-290.
© 1997 *Kluwer Academic Publishers. Printed in the Netherlands.*

Need for International Efforts and Role of International Organizations

Basin states in developing regions may lack the capacity to develop and manage their own shared water resources. The international community may provide both technical and financial support to develop and implement an integrated water resources management scheme. Through cooperation, states may obtain aid that might not otherwise be available to them [3].

Overcoming institutional barriers, either within a basin country or among riparians, is not easy. Institutions that manage, such as river-basin authorities, often involve only one ministry in a basin country and the decisions of such an institution may not be conveyed to the central decision-making mechanism of the basin country. Specific treaties or agreements are needed to codify the responsibilities of participating nations and their facilitating agencies.

The lessons of experience with agreements and joint actions among riparians, such as the World Bank's difficult but successful nine-year effort to facilitate the 1960 Indus Water Treaty between India and Pakistan, suggest that external assistance and encouragement are valuable and sometimes essential ingredients in establishing international water agreements. Where the basic institutional framework exists, international agencies should provide support and encouragement. International agencies can also assist riparians in developing and managing water resources and in facilitating the implementation of treaties.

The three main objectives of international efforts should be (a) to help riparian countries address problems with international water resources, (b) to release priority development activities that are held hostage by disputes over shared watercourses, and (c) to reduce inefficiencies in the use and development of scarce water resources caused by the lack of cooperative planning and development. Since no single international organization commands all the skills, experience, or resources necessary to achieve the needed cooperation, collaborative efforts among potential donors, international organizations, and NGOs could promote the sound management of international waters [5].

Indus Water Treaty: A World Bank Success Story

The World Bank's Policy Paper on Water Resources Management notes that "External assistance and encouragement are valuable and sometimes essential ingredients in establishing international water agreements" with regard to the Indus Water Treaty agreed on by India and Pakistan in 1960 [5].

The Indus River basin was divided between India and Pakistan, following the independence of these countries in 1947. Some of Pakistan's irrigated farmlands were then fed by rivers on the Indian side of their international border. An agreement on the use of shared water resources was needed, but the two riparians failed to come to an agreement through bilateral talks. The World Bank then offered to help resolve the dispute, and both India and Pakistan agreed [1]. However, in the course of the negotiation, the idea of "integrated management" of the Indus River basin by riparian countries (namely Pakistan and India) was abandoned. The treaty agreed on the division of the catchment (with six rivers) by the two countries (three rivers each) so that each country could manage "its own rivers" as it wished.

The Indus Water Treaty can be regarded as a success story of the World Bank, from the viewpoint of the World Bank acting in the role of mediator, though it was achieved at the sacrifice of the integrated approach toward which the negotiating process was originally aimed.

Why Have There Not Been More World Bank Success Stories in the Past 35 Years?

A World Bank technical paper on international inland waters [2] noted that "with the notable exception of the Indus Water Treaty, the Bank has made only limited direct intervention in international water affairs." It also mentioned that since the signing of the Indus Treaty, the World Bank had been ready to offer its good offices to resolve development problems of international rivers such as the Ganges, Nile, Tigris and Euphrates. In no case, however, has it played as proactive a role as it had for the Indus situtation, that is, to stimulate the riparians concerned to accept its good offices.

The lack of another World Bank success story over the past 35 years (after the Indus Water Treaty) may be attributed to various factors. Some critics suggest that it is due to the lack of strong will and persistence, as had been shown by Eugene Black who was then the World Bank President. Others suggest that a more proactive approach would be required for the World Bank to have another success story. Though these views have an element of truth in them, a most important factor is that the circumstances, in the field of overseas development aid, have changed and that today the World Bank is simply one of several funding agencies. Back in the 1950s, the World Bank was one of very few agencies to give assistance to the developing world, and could act as the coordinator among these few donor agencies and countries.

Numerous agencies are now in the field of overseas development assistance, and the lack of coordination among these agencies is obvious and commonplace. Thus, the World Bank may no longer be able to take a leading role in dealing with international water bodies, in the same way that it did for the Indus River basin.

Can International Organizations Still Do Something Effective for the Caspian Sea?

The above-mentioned difficulty does not necessarily mean that international organizations can do little for international water bodies nowadays. What seems to be difficult is to undertake "direct intervention" as the World Bank did for the Indus River basin. International organizations have made several contributions by way of "indirect intervention" with some success stories (and, to be honest, many failures) in the last few decades. The following paragraphs briefly highlight success stories and failures. It is hoped that from such cases we may learn what international organizations could do, as well as what they should not do for the Caspian Sea.

The Role the World Bank Played in the New Nile Water Treaty in 1959

In 1959, Egypt (then the United Arab Republic) and Sudan supported a new agreement on the use of water resources in the Nile River. It superseded the old (1929) agreement between the same two riparians. Both Egypt and Sudan needed this new agreement for their then-planned High Aswan Dam and Roseires Dam (on the Blue Nile), respectively.

The High Aswan Dam was financed by the U.S.S.R. after the cancellation of financial assistance that had been pledged by the U.S.A., U.K. and the World Bank. The High Aswan Dam is by no means a success story for the World Bank in this context. However, for the Roseires Dam project in Sudan, the World Bank was THE major donor agency and it requested that the Sudanese Government sign an agreement with Egypt about sharing Nile River water. The Sudanese Government acted accordingly, and the two riparian countries commenced negotiation towards a new agreement. The role of the World Bank was rather indirect in that it did not act as the intermediate body as it had for the Indus River Treaty. However, it played a major role in getting the two basin countries to agree on the new treaty.

Other International Water Body "Success Stories"

The Mekong Committee (now the Mekong River Commission) has maintained the basin's hydrological and meteorological monitoring stations for more than 30 years. It is one of the well monitored international river basins. Taking into consideration that the lack (or paucity) of harmonized hydrological data in an international basin has often been the source of conflict among basin states (e.g., the lengthy dispute between India and Bangladesh over the Farakka Barrage on the Ganges River, and the early dispute over the Indus River between Pakistan and India), the Mekong's "standard" hydrological dataset is a great asset for basin countries in the negotiating of development projects. The success of the Mekong Committee can be attributed to the support and guidance provided by the UNDP and UN-ECAP over a period of more than three decades.

Recently, the UNDP took a leading role in the creation of the "new" Mekong River Commission through numerous negotiations in the 1992-1994 period. In early 1992, the Mekong Committee (then the Interim Mekong Committee) was on the verge of demise due to a conflict of interest between Thailand and other member states, in particular Vietnam. Thailand, as an upstream country, planned to divert water from the Mekong River to its territory for development of irrigated farmland. Thailand hinted that it would withdraw from the (then Interim) Mekong Committee if other "downstream" member states intended to veto its plan, according to the original agreement of the Mekong Committee. The UNDP, which had been one of the major donors for the Mekong Committee, became involved. Through a number of meetings and consultations, both formal and informal, the UNDP tried to keep Thailand within the framework of the Mekong Committee. It was "only the pressure from the donors (including the UNDP) that saved the Mekong Committee from break-down" [4]. The member countries, including Thailand, adopted the new agreement in 1995. This new agreement is less binding for riparian countries, in the sense that a member country may not "veto" the conduct of another riparian state. The involvement of the UNDP in the creation of the new agreement is yet another example of a rather direct and leading role taken by an international organization.

The European Union (EU) was instrumental in the resolution of a dispute over two dams in the Danube River basin between Hungary and Slovakia. These two countries had an agreement in the 1970s to build two dams (one for each country) along the Danube River. After the change of its political regime in 1989, Hungary decided to abandon construction of its dam, due mainly to objections expressed by environmentalists within the country, and requested Slovakia to do the same. Slovakia objected to the request and continued construction of its dam. After the failure of bilateral negotiations, the European Union acted as mediator, and an agreement was reached so that both basin countries would follow the "verdict" of the International Court of Justice, to be issued in 1997. The bottom line of this case is that both of these countries wish to be member states of the EU and that they are obliged to resolve the conflict. The EU was thus in a position to be influential, though rather indirectly, for basin countries.

The Zambezi Action Plan as an Agreement on Environmental Issues

International organizations have been involved in the environmental aspects of international water bodies in only a few exceptional cases. The Zambezi Action Plan is one such rare example.

The Zambezi Action Plan was agreed upon in 1987 by the six countries of the Zambezi River basin in Southern Africa. It was the first international agreement which aimed at dealing with environmental aspects of an international water body. The United Nations Environment Programme (UNEP) took the leading role in the formulation of the Action Plan, by acting as the secretariat during its preparation. UNEP met most of the cost for its preparation. The Diagnostic Study on the environmental aspects of the basin was developed between 1985 and 1987. The Action Plan was elaborated after the Diagnostic Study.

However, after its adoption, few projects listed in the Action Plan were implemented. The Zambezi Action Plan has so far had very limited success, because of the following reasons. First of all, the agreement was sought mainly to meet political needs. The Zambezi River basin was designated as the priority area nominated in the Cairo Action Plan, a decision made by African ministers on the environment adopted in Cairo in 1985. The riparian countries were obliged to show their will to promote environmental activities within the basin. They were, however, not highly motivated to work on the substance of the agreement. The lack of collaboration within donor countries, not among basin countries, was another reason. The concept in the Action Plan was too general, not identifying specific projects to promote donors' interest. Institutional arrangements among basin states were also insufficient for implementation of the Action Plan.

The lesson gained from the Zambezi Action Plan is that the following aspects are indispensable for an international agreement to be implemented with a high chance for success: (i) a proper motivation for an agreement, (ii) coordination among basin countries, (iii) promotion of donors' interest, and (iv) appropriate institutional arrangements for implementation.

Discussion

As mentioned above, international organizations could still have roles to play with regard to international water bodies, although they may not be as dramatic as the role of the World Bank in the Indus River basin in the 1950s. Examples from some past "success stories" could give guidance to those concerned about the Caspian Sea basin.

Riparian countries might argue endlessly over the reliability of data on which any management plan should be based, as was the case in the Indus and Ganges River basins. The "old Mekong model" of the Mekong Committee could be a good model, in terms of establishing a mutually agreeable monitoring system for the wide distribution of data, and implementation of a long-term scheme. Considering the fact that no universally acceptable database exists among Caspian Sea basin countries, establishment of such a system should have the utmost importance.

An international organization can act in an indirect but essential way as an influential third party, as shown in the "Nile model" and "Danube model". Based on these experiences, an international organization should have an effective "carrot and stick" policy when dealing with riparian countries to encourage the development of an agreement on the use of their shared water resources. In the Nile case, the "carrot" was financial aid for Sudan and Egypt from the World Bank and the U.S.S.R., respectively. The same situation occurred with the membership of the EU in the Danube case. The same approach may be applied to the Caspian Sea basin, only if an international organization has a "carrot" which every basin country wishes to obtain at the expense of putting aside their differences to achieve an agreement.

The "Indus model" or the "new Mekong model," in which an international organization may serve as an intermediary could still be used, provided that riparian countries *and* donor agencies are willing to be coordinated by international organizations. For this model to be successful, a prerequisite is that the specific international organization should be trusted and empowered not only by basin countries but also by donor countries and agencies as well. This condition is rather difficult to meet, because few, if any, donor countries or agencies are willing to be coordinated by others. This scheme could work only if an international organization was able to persuade other donors that the sole intermediary would provide them with many benefits which they may otherwise not get. It is yet to be seen if an international organization can establish such a scheme in the Caspian Sea basin, given that various donor countries and agencies presently tend to take a country, rather than a basin, approach in their aid operations.

As a final comment, having basin countries adopt an agreement does not guarantee the effective implementation of the agreement. The riparian countries of the Caspian Sea may see the "Zambezi model" from this point of view. Unless basin countries are really willing to implement the agreed-upon scheme by sharing their human and financial resources, nothing of substance will happen. Riparian countries, for their part, must make their proposed management scheme attractive, substantial, project-oriented and realistic in order to secure support from donors in the implementation phase.

References

1. Barrett, S. (1994): Conflict and Cooperation in Managing International Water Resources, pp. 11-12, *Policy Research Working Paper 1303*, World Bank, Washington D.C.

2. Kirmani, S. and Rangeley, R. (1994): *International Inland Waters - Concepts for a More Active World Bank Role*, pp. 3-17, World Bank Technical Paper 239, World Bank, Washington DC.

3. LeMarquand, D.G. (1981): International Action for International Rivers. *Water International*, **6**, 147-151.

4. Ojendal, J. (1995): Mainland Southeast Asia: Co-operation or Conflict over Water?, in L. Ohlsson (ed.) *Hydropolitics*, 149-177, Zed Books, London.

5. World Bank (1993): Water Resources Management, World Bank Policy Paper, World Bank, Washington D.C., 63 pp.

6. World Bank (1994): *A Guide to the Formulation of Water Resources Strategy* - World Bank Technical Paper Number 263 -. World Bank, Washington, D.C.

Appendix

List of Participants by Country
NATO Advanced Research Workshop
Scientific, Environmental, and Political Issues in the Circum-Caspian Region
13-16 May 1996
Moscow, Russia

Azerbaijan:

Arif Mekhtiyev
Director, National Aerospace Agency
Baku, Azerbaijan
Tel: 99412-629387
Fax: 99412-621738

Canada:

Amir Badakhshan, Professor
Dept. of Chemical & Petroleum Engineering
University of Calgary
Tel: 403-220-4799
Fax: 403-282-4899

Estonia:

Sergei Preis
Tallinn Technical University
Tallinn, Estonia
Tel: 7-3722-532115
Fax: 7-3722-532446
spreis@edu.ttu.ee

Hungary:

Pavel Szerbin
Head, Division of Environmental Radiohygiene
Joliet-Curie National Research Institute
POB 101, Budapest 1775, Hungary
Tel: 36-1-227-0368
Fax: 36-1-226-6531
pavel@hp.osski.hu

Iran:

Abbasgholi Jahani
General Director of Water Planning Bureau
Ministry of Energy
81 North Felestin Avenue, Tehran, Iran
Fax: 98-21-880-1555

Jalal Shayegan
Sharif University of Technology
Tehran, Iran
Fax: 98-21-601-2983

Japan:

Mikiyasu Nakayama
2-1-2-1202 Higashinakano
Nakano-ku, Tokyo 164, Japan
Tel: 81-3-33361-5385
Fax: 81-3-33361-5389
nakayama@utsunomiya-u.ac.jp

Kenya:

Gerhart Schneider
UN Environment Programme
Nairobi, Kenya
Tel: 254-2-621234
Fax: 254-2-226886
gerhart.schneider@unep.no

Russia:

George Golitsyn
Institute for Atmospheric Physics
Pyzhevsky per. 3
Moscow, Russia
Fax: 7-095-2649187

G. Golubev
Faculty of Geography
Moscow State University
Vorobiovy Gory
Moscow, Russia
Fax: 7-095-9328836

I. Granberg
Institute of Atmospheric Physics
Pyzhevsky per. 3
Moscow, Russia
Fax: 7-095-2331652

Sergei Metveyev
c/o I.S. Zonn, UNEPCOM
11, Novy Arbat Str., Bldg. 1, Room 1819
121019 Moscow, Russia
Tel/Fax: 7-095-267-4797

Nina Novikova
Institute of Water Problems
Russian Academy of Sciences
10, Novaya Basmannaya str.
Moscow 107078, Russia
Tel: 7-095-265-9565
Fax: 7-095-265-1887
novikova@iwapr.msk.su

G. Voropayev
Scientific Coordination Center
Kadashevsky per. 10
113035 Moscow, Russia
Fax: 7-095-2316841

Z. Zalibekov
PriCaspian Institute of Bioresources
367025 Makhachkala, Mt. Gadgiev st. 45
Republic of Dagestan, Russia

Igor Zonn, Co-Director
Vice President, UNEPCOM
11, Novy Arbat Str., Bldg. 1, Room 1819
Moscow 121019, Russia
Fax: 7-095-135-6159

Sweden:
João Morais
Programme Officer for Social Sciences
IGBP Secretariat
The Royal Swedish Academy of Sciences
Box 50005, S-10405, Stockholm, Sweden
Tel: 46-8-166448
Fax: 46-8-166405
morais@igbp.kva.se

Turkmenistan:
A. Babayev
Director, Desert Resarch Institute
Gogol St., 15
Ashgabat, Turkmenistan
Fax: 253716

Turkey:
Münir Özturk
Centre for Environmental Studies
Bornova, Izmir, Turkey
Tel: 90-232-388-0110
Fax: 90-232-388-1036

Secretariat: **D. Jan Stewart**
ESIG/NCAR

Ilter Turan
Department of Political Science
Koc University
Istinye 80860 Istanbul, Turkey
Tel: 90-212-229-3006
Fax: 90-212-229-3602
ituran@ku.edu.tr

United Kingdom:
Tatiana Saiko
Department of Geographical Sciences
The University of Plymouth
Drake Circus, Plymouth
Devon, UK PL4 8AA
Tel: 44-1752-233053
Fax: 44-1752-233054
tsaiko@plymouth.ac.uk

Sergei V. Vinogradov
Center of Petroleum and Mineral
Law and Policy
University of Dundee
Dundee DD1 4HN Scotland, UK
Tel/Fax: 44-1382-552294
svvinogradov@its.dundee.ac.uk

United States:
Michael H. Glantz, Co-Director
Program Director
Environmental & Societal Impacts Group
National Center for Atmospheric Research
PO Box 3000
Boulder, CO USA 80307
Tel: 303-497-8119
Fax: 303-497-8125
glantz@ucar.edu

John Kineman
Natl Oceanic & Atmospheric Administration
National Geophysical Data Center
325 Broadway
Boulder, CO USA 80303-3328
Tel: 303-497-6900
Fax: 303-497-6513
jjk@ngdc.noaa.gov

Uzbekistan:
V. Sokolov
Interstate Coordination Water Commission
B-11, Karasu-4
Tashkent, Uzbekistan
Tel: 7-3712-650955
Fax: 7-3712-650558

BIOGRAPHICAL STATEMENTS OF PARTICIPANTS

NATO Advanced Research Workshop
Scientific, Environmental, and Political Issues in the Cirrum-Caspian Region

Agadjan G. Babaev is a Doctor of Geography, Professor, and a Member of the Turkmen Academy of Science, he is also a corresponding member of the Russian Academy of Science and a Director of the Desert Research Institute of the Turkmen Academy of Science. Dr. Babaev is Editor-in-Chief of the international journal, *Problems of Desert Development* (in Russian and English).

Dr. Babaev is a Foreign Member of the Academy of Natural Science and an honorary member of the Islamic Academy of Science. He is the USSR State Prize Winner, Winner of the Karpinski International (Germany) and J. Piel Prizes (USA). He has published over 300 research and scientific popular works, including 12 monographs.

Amir Badakhshan is a Professor, Ph.D., Birmingham University, U.K. Who was appointed to the University of Calgary in 1982. He teaches Petroleum Production Engineering, Environmental Engineering Aspects of Water Pollution, Environmental Engineering Aspects of Air Pollution, and supervises Engineering Design Projects of final-year students. His major areas of research include Environmental Engineering Aspects of Water Treatment, Enhanced Oil Recovery by Chemical Flooding, Liquid-liquid Equilibrium of Ternary and Quaternary Systems and Clay Stabilization in Petroleum Reservoirs. Dr. Badakhshan has authored and co-authored more than 90 papers published in refereed journals and presented at national and international conferences.

He has been Associate Dean and Assistant Dean of Engineering at the University of Calgary and is a consultant to the United Nations U.N. Development Program. Dr. Badakhshan has been a Visiting Professor (one term per year) at University Petroleum Technology and Sharif University of Technology in Iran since 1990, teaching Advanced Environmental Engineering to Post-Graduate students and supervising Graduate students in research topics and thesis preparation.

Yurij S. Chuikov is a Doctor of Biological Sciences, member of the Russian Academy of Ecology. He graduated from the Astrakhan Technical Institute of Fishery Industry and Economy in ichthyology. He also took a post-graduate course at Moscow State University in zoology.

From 1973 he worked in the Astrakhan State Biosphere Reserve as a junior senior scientific worker, deputy director on scientific work. He received a Ph.D. from Moscow State University and a Doctorate from the Institute of Zoology in St. Petersburg, Russia. He is the author of 185 scientific papers on the problems of zoology, hydrobiology, ecology of aquatic and earth ecosystems published in Russia and abroad. He participated in a number of international programs, conducted by the World Wildlife Fund and the International Bureau on Conservation and Study of Wetlands, of joint Russian-Netherlands programs on the Volga Delta, and has participated in numerous international and Russian conferences and symposia.

At present he is the Chairman of the Committee of Ecology and Natural Resources in the Ministry of Ecology and Natural Resources of the Russian Federation representing Astrakhan. He heads several regional research programs for the establishment of a system to protect natural areas and to study the impacts of pollution on natural ecosystems.

Michael H. Glantz is a Senior Scientist in the Environmental and Societal Impacts Group, a program at the National Center for Atmospheric Research (NCAR). He is interested in how climate affects society and how society affects climate, especially how the interaction between climate anomalies and human activities affect quality of life issues. His research relates to African drought and desertification and food production problems and prospects; societal impacts of climate anomalies related to El Niño events, and the use of El Niño-related teleconnections to forecast these impacts; to developing methods of forecasting possible societal responses to the regional impacts of climate change; and the use of climate-related information for economic development. He has also coordinated joint research in the Central Asian Republics of the ex-USSR.

He received his B.S. in Metallurgical Engineering (1961) and an M.A. in Political Science (1963) from the University of Pennsylvania. After some years in industry (Westinghouse, Ford), he returned to the University of Pennsylvania and received a Ph.D. in Political Science in 1970. He has taught at the University of Pennsylvania, Lafayette College, and Swarthmore College. In 1974 he joined NCAR as a postdoctoral fellow, and in 1983 he was the first social scientist to became a Senior Scientist at that institution. He has edited several books and is the author of numerous articles on issues related to climate, environment, and policy. His most recent publication is *Currents of Change: El Niño's Impact on Climate and Society*, published in the fall of 1996 by Cambridge University Press.

George S. Golitsyn graduated in 1958 from the Physical Science Faculty of Moscow State University. Since February 1958 he has worked in the Institute of Atmospheric Physics of the USSR (now Russian) Academy of Science. His research is on turbulence theory, wave propagation and generation in the atmosphere, dynamics of the atmosphere, including the planetary atmosphere, convection theory and experiments with application to the boundary layers, air-sea interactions, mantle convection, convection on astrophysical objects, climate and climate change theory, and the theory of earthquakes.

He has authored or co-authored more than 200 scientific papers and 5 monographs. He has been director of the Oboukhov Institute of Atmospheric Physics since 1990, a member of the Presidium of the Russian Academy of Science, Chairman of the Russian Interagency Scientific Council on the Problems of the Caspian Sea, and Chief Editor of the magazine *Izvestia-Atmospheric and Oceanic Physics*. He is also a member (1982-1987, 1991-1996) of the Joint Scientific Committee which advises the World Climate Research Programme (WCRP) and has had several other national and international positions.

Genady N. Golubev is Professor and Head of the Department of Physical Geography of the World and Geoecology, Faculty of Geography, Moscow State University. His interests and expertise are in strategies of environmental management and sustainable development; global environmental change research; state of the environment and related strategies; water resources, management, water-related environmental problems; education and training in the area of environment and development; environmental problems of agriculture; and problems of mountains, deserts and desertification.

He received his Ph.D. in Hydrology in 1963 and D.Sc. in Hydrology in 1974 and became Professor of World Physical Geography and Geoecology in 1992. All his degrees are from Moscow State University, with which he has been affiliated all of his professional life. Between 1976 and 1981 he was research scholar and task leader at the International Institute for Applied Systems Analysis (IIASA) in Laxenburg, Austria. From 1981 through 1989 he served as an Assistant Secretary General of the United Nations and as an Assistant Executive Director of UNEP (United Nations Environment Programme). In 1992-1993 he served as Assistant Director General at the World Conservation Union (IUCN). He has been and continues to be a member of numerous national and international committees, editorial boards and advisory councils related to environmental issues. Dr. Golubev has authored more than 150 publications, mostly in Russian and English, including 13 books authored, co-authored, edited and co-edited covering the issues mentioned above.

Abbasgholi Jahani is the General Director of the Water Planning Bureau of the Ministry of Energy in Tehran, Iran. He is a member of the Commission on Hydrology of the World Meteorological Organization. He is also Iran's representative for the intergovernmental council of UNESCO. He is a member of the national committees for IHP, ICID and ICOLD. He represents Iran as a member of the working group for the water balance study of the Caspian Sea.

John Kineman currently works as a Physical and Ecologist in Boulder, Colorado, where he leads an *Ecosystems Data and Information Group* within the NOAA National Geophysical Data Center. He received his Bachelors Degree in Earth Physics in 1972 from the University of California, Los Angeles, and a Masters of Basic Science in multi-disciplinary environmental sciences in 1979 from the University of Colorado, Boulder. He pursued further graduate work toward the Ph.D. in Ecology at the University of California, Davis, before returning to Government Service in his present field.

Mr. Kineman gained wide experience through a series of government positions, including: conducting geophysical surveys and environmental assessments in the NOAA Commissioned Corps (reaching the rank of Lieutenant Commander); research planning as a Game Warden in the Kenya Wildlife Service; independent field work in wildlife conservation research in Africa; and work in ecosystem characterization and database management leading to his present position. Mr. Kineman has numerous publications on the integration of ecosystems data using Geographical Information Systems (GIS), and improvements to the delivery of Ecological Characterization information using computer methods and digital databases. He has also published on diverse topics including wildlife conservation, the philosophy of science, and evolution.

Arif Sh. Mekhtiev, after graduating from the Moscow Electrotechnical University in 1957, began his scientific activity as a researcher in Academy of Science of Azerbaijan. In 1969 he received his Candidate of Science degree and in 1984, Doctor of Science. In 1975 he joined the organization later named the Azerbaijan National Aerospace Agency (ANASA) as a Vice Director responsible for scientific programs, including a program for studying Caspian Sea problems. Since 1991 he has been the General Director of ANASA.

Dr. Mekhtiev 's main scientific field of interest is remote sensing methods and techniques, and its application in solving various ecological problems. He is an author of more than 100 scientific publications on different aspects of space applications.

João M.F. Morais is Program Officer for Social Sciences at the International Geosphere-Biosphere Programme (IGBP) Secretariat. Professor Morais is a Portuguese citizen who received his Ph.D. in Archaeology from Oxford University in the UK, after studies at Lourenço Marques University in Mozambique, where he directed the Department of Archaeology and Anthropology. He has worked as a principal researcher at the Tropical Research Institute (IICT) and at the Universidade Lusofona de Humanidades e Technologias in Lisbon, Portugal. His academic interests are in African archaeology, early farming systems, past global changes and environmental archaeology, and interaction processes in the natural and social sciences.

Dr. Morais's work addresses issues on the linkages between the natural and social sciences in the study of global change. He acts as the primary focal point for Land-Use/Cover Change (LUCC), for Past Global Changes (PAGES), and for the Land-Ocean Interactions in the Coastal Zone (LOICZ). He provides liaison between the IGBP and IHDP, the European Network for Research in Global Change Program of the European Community (ENRICH), and the international community of Social Scientists.

Mikiyasu Nakayama is an Associate Professor of the Faculty of Agriculture, Utsunomiya University, Japan. He is interested in environmental monitoring and management of river and lake basins. His research subjects include the application of satellite remote sensing data for environmental monitoring of lake basins, the use of Geographical Information System (GIS) for environmental management of river and lake basins, environmental impact assessment methodologies applicable for involuntary resettlement due to dam construction, and the involvement of international organizations in management of international water bodies.

He received his B.A. (1980), M.Sc. (1982) and Ph.D. (1986) from the University of Tokyo. He served as program officer in the United Nations Environment Programme (UNEP) between 1986 and 1989. In UNEP, he participated in projects on international water bodies such as the Zambezi River, Lake Chad, and the Mekong River. From 1989 to present, he has been teaching courses on water resources management at Utsunomiya University. He has been also serving as an advisor and an expert for several United Nations organizations (UNEP, UNCHS, UNCRD, and UNU), as well as for non-governmental organizations. He participated in UNEP's environmental management project for the Aral Sea between 1990 and 1992. From 1994 to 1996, he was "on loan" to the North African Department of the World Bank to deal with water resources management projects in Morocco, Tunisia and Iran. He has published eleven books along with other authors and about 30 scientific papers on water resources management.

Münir Öztürk is the Director of the Center for Environmental Studies, Faculty of Science, Ege University. He received an M.Sc. (Kashmir University, 1964); Ph.D. (Ege University, 1970), D.Sc. (Ege University, 1975). He has been a Professor at Ege University since 1980. He is a member of national and international societies as well as committees on Ecology, Ecosystem Health, Biology and Environmental Protection. He teaches World Ecosystems, Plant Ecology, Pollution Impacts, Soil-Plant Relationships, and Coastal Management. He has nearly 200 publications in English and Turkish. He has edited two symposia volumes in English. He has spent some time in Germany and Japan as an "Alexander-von-Humboldt" fellow and as a "Japan Society for Promotion of Science" fellow, respectively.

Sergei Preis is a Senior Scientist of the Department of Chemical Engineering, Tallinn Technical University (TTU), Estonia. His scientific interest is in the field of oxidation technologies in water and wastewater treatment. His research relates to photocatalytical oxidation on non-biodegradable toxic substances such as phenols, amino and nitrocompounds and oil spills. He supervises a group of M.Sc-students, supported by the Estonian Science Foundation. He also participates in collaboration between TTU and Lappeenranta University of Technology in Finland.

He received a Diploma of Chemical Engineer (1981) from TTU. After two years in the military, he returned to TTU and received a Candidate of Science (Ph.D.) In Water Supply and Sewerage in 1988. He has been involved in research in oxidation technologies for water/wastewater treatment and has taught at TTU in the Department of Chemical Engineering. From 1989 to 1992 he was the Head of TTU's Laboratory of Chemical Engineering and Environmental Protection. Since 1992 he has been a Senior Scientist in the Department of Chemical Engineering at TTU. He visited Imperial College of Science, Technology and Medicine, London, UK and Lappeenranta University of Technology, Lappeenranta, Finland, for ten months each as a researcher.

Tatyana A. Saiko is a Lecturer in Geography at the Department of Geographical Science, Faculty of Science, University of Plymouth in the UK. She teaches courses on topics related to arid environments, to problems of desertification and to environmental crises in East-Central Europe and the former Soviet Union (FSU). Her research focuses on these issues, with special interest in environmental problems of arid regions in the FSU.

She received her M.Sc. in Geography (1972) at Moscow State University, Russia, and her Ph.D. in Environmental Protection and Management of Natural Resources (1986) at the Institute of Desert Research in Ashgabat, Turkmenistan. Her thesis was on the problems of anthropogenic desertification in the Sudano-Sahelian region of Africa. After graduation from the university she worked as a researcher at Moscow State University, specializing in remote sensing of the environment. In 1980 she began to work for the Center for International Projects (CIP) of the USSR Commission for UNEP in Moscow. There she became CIP's Deputy Coordinator of the Terrestrial Ecosystems Division, where she was responsible for international training and research activities on desertification-related issues of concern to UNEP, both in the former Soviet republics and in developing countries such as Botswana and Mali. Since 1993, she has been teaching at the University of Plymouth, United Kingdom.

Professor Saiko has participated in 28 international conferences, workshops and training courses. She is the author of numerous articles related to desertification and its control and environmental problems of the FSU and the new democracies of Europe.

Jalal Shayegan, Ph.D., is a Professor at the Sharif University of Technology, Tehran, Iran. He is a member and the head of the Environmental Division of Iran's Academy of Science. He received his B.Sc. from Tehran University, and his M.Sc. and Ph.D. from Aston University in Birmingham, UK. He has published more than 50 papers and supervises more than 20 post-graduate students. He teaches subjects in the fields of Environmental Energy, Water Pollution Control and Anaerobic Treatment. The positions that Dr. Shayegan has held include Head of Department of Chemical Engineering, Sharif University, Tehran; Deputy Head of the Biochemical and Bioenvironmental Research Center at Sharif University, Head of Palab Institute for Water Research, Tehran; Deputy Head for the Organization of Mine and Petroleum Products, Tehran; Member of Energy Research Center University of Wisconsin, Madison, Wisconsin, USA; and Visiting Professor at the University of Toronto, Canada.

Vadim I. Sokolov is the Head of the Water Management Division of the Scientific-Information Center of Interstate Coordination, Water Commission for Central Asia. He received his Ph.D. in hydrology and water management from the Institute of Geography in the Russian Academy of Sciences in 1991. He is also an engineer in hydrotechnical construction, a degree he received from the Tashkent Institute of Irrigation in 1981.

Pável Szerbin is The Head of the Division of Environmental Radiohygiene of the "Frédéric Joliot-Curie" National Research Institute for Radiobiology and Radiohygiene, Budapest, Hungary. He received his M.Sc. in Radiobiology from Moscow State University (1978). He then worked at the Department of Ecology and Taxonomy of the Eötvös Lóránd Science University in Budapest. After military service he returned to the USSR and received his Ph.D. in Radiobiology at the Department of Roentgenology at the Leningrad Veterinary Academy (1985). Since then, he has been at his present institute. He is a member of different national and international scientific bodies (ESRB, IUR, etc.). He and his division have been involved in several national research programs and international joint research projects. The results of these studies have been published in different scientific journals.

Gerhart Schneider holds an M.Sc. in Agro-biology, and wrote his thesis on clay soils in Southwestern Germany. After working as a soil scientist in West Africa in 1987, he joined the Environmental Office of the city administration of Stuttgart, Germany. Since 1989, he has been with the United Nations Environment Programme in Nairobi, Kenya. Highlights of his work include: overseeing research projects on the biogeochemical cycles of phosphorus and carbon; work on problems of management of international water bodies, especially Lake Victoria, and rivers such as the Nile and the Mekong.

Ilter Turan received his B.A. in Government from Oberlin College, Ohio 1962, an M.A. in Government from Columbia University in 1964 and a Ph.D. from the Faculty of Economics of Istanbul University in 1966. He has been Assistant Professor of Political Science (1966-1970), Associate Professor of Political Science (1970-1976) and Professor of Political Science (1976-1984) and Chair of Political Science of the Faculty of Economics of Istanbul University. In 1984, he became a member of the Faculty of Political Science at Istanbul University where he served as the Chair of the Department of International Relations. Since 1993, he has been a member of the Department of International Relations at Koç University of Istanbul.

Dr. Turan has held visiting appointments at various American institutions including the universities of Iowa, Wisconsin (Madison), Arizona, Kentucky, and California (Berkeley). He is the author of many books and articles in Turkish and in English.

Sergei V. Vinogradov, Doctor of Law, Moscow State University, is a Senior Research Fellow at the Centre for Energy, Petroleum and Mineral Law and Policy, University of Dundee, Scotland, UK. His main areas of interest include the law of the sea, international environmental law, the law of international watercourses, and environmental and natural resource law of Russia and the Commonwealth of Independent States. Currently, he is studying the regulation of offshore petroleum activities and the legal status and regime of the Caspian Sea. His works have been published in Russia, Germany, USA, UK, The Netherlands, Canada, and Bulgaria. He is a member of the Water Resources Committee of the International Law Association and the Commission on Environmental Law of the IUCN/World Conservation Union.

Grigory V. Voropaev graduated in 1995 from the Moscow Institute of water-economy engineering. He is an Engineer of water economy, hydrologist, and a Doctor of technical sciences. He is a member of the Russian Academy of Sciences (RAS) and of the Academy of Water Economy Sciences in Russia.

From 1954 to 1970, Dr. Voropaev worked on the arid regions of the Middle Asia and Kazakstan investigating water resources, water utilization and reclamation of land in the region. Since 1971, he has worked in the USSR Academy of Sciences. From 1975 to 1988, he served as the Director of the Institute of Water Problems. At present he is the Director of the Scientific Coordination Center (SCC). He was the editor-in-chief and, now, a member of the Editorial Board of the journal of the Russian Academy of Sciences, *Water Resources*, which has been published in the USA in English since 1979.

Since 1976, Dr. Voropaev has been the Chairman of the Scientific Council of the RAS on the Complex Problems of the Caspian Sea and its Basin and also Co-chairman of the Joint Russian-Iranian Working Group on the Caspian Sea that executes the program of scientific and technical cooperation between the Islamic Republic of Iran and the Russian Federation. He participated in the development and realization of large projects related to the USSR water problems. He has published more than 300 scientific papers.

Zalibek G. Zalibekov is a Doctor of Biological Science, Professor, and Director of the PreCaspian Institute of Biological Resources in the Russian Academy of Science (RAS), chairman of the section of the Science Council of the RAS, "Problems of the Study of Arid Ecosystems and the Combat Against Desertification," and chief editor of the magazine, *Arid Ecosystems*. His interests in soil science research cover the anthropogenic influence on soils development. Lately he has been working on the biological issues of the Caspian Sea. He is the author of 120 articles and 3 books.

Igor S. Zonn is Doctor of Geography, an Academic Advisor of the Russian Engineering Academy, and a corresponding member of the Russian Academy of Natural Science. He is the Vice President of the Russian organization UNEPCOM, Head of a Department in the Engineering Center for Water Resources, Ecology and Reclamation, a member of the Editorial board of the international magazine *Problems of Desert Development*, and a member of the International Advisory Committee, "Desert Technology." He is interested in how society affects water resources, in the environmental consequences of reclamation land, in conflict, climate change, problems of desertification, political ecology, and international water usage.

From 1991 to 1995, Dr. Zonn worked on Aral Sea problems with scientists and engineers from the Japanese non-governmental organization, the Global Infrastructure Fund (GIF), and several Japanese universities. During the last five years, he has worked with ESIG/NCAR, USA. He is the author of numerous articles, monographs, book chapters and technical reports.

INDEX